BASIC TECHNICAL COLLEGE MATHEMATICS

BASIC TECHNICAL COLLEGE MATHEMATICS

Second Edition

Len Mrachek
Charles Komschlies

Regents/Prentice Hall
Englewood Cliffs, New Jersey 07632

Library of Congress Cataloging-in-Publication Data

Mrachek, Len
    Basic technical college mathematics / Leonard Mrachek, Charles Komschlies. —2nd ed.
      p. cm.
    Rev. ed. of: Technical-vocational mathematics. 1st ed. 1978.
    Includes index.
    ISBN 0-13-891995-X
    1. Mathematics.   I. Komschlies, Charles G.
II. Mrachek, Len. Technical-vocational mathematics.
III. Title
QA39.2M7   1993
513'.14—dc20   92-33729
                                                                     CIP

Acquisitions editor: Frank Burrows
Editorial/production supervision and
   interior design: Tally Morgan, WordCrafters Editorial Services, Inc.
Cover design: Marianne Frasco
Prepress buyer: Ilene Sanford
Manufacturing buyer: Ed O'Dougherty

Cover: Meshing gears, FPG; circuit board, Otto Rogge/The Stock Market; construction workers, FPG.

© 1993, 1978 by Regents/Prentice Hall, Inc.
A Division of Simon & Schuster
Englewood Cliffs, New Jersey 07632

All rights reserved. No part of this book may be
reproduced, in any form or by any means,
without permission in writing from the publisher.

Printed in the United States of America
10 9 8 7 6 5 4 3 2 1

ISBN 0-13-891995-X

Prentice-Hall International (UK) Limited, *London*
Prentice-Hall of Australia Pty. Limited, *Sydney*
Prentice-Hall Canada Inc., *Toronto*
Prentice-Hall Hispanoamericana, S.A., *Mexico*
Prentice-Hall of India Private Limited, *New Delhi*
Prentice-Hall of Japan, Inc., *Tokyo*
Simon & Schuster Asia Pte. Ltd., *Singapore*
Editora Prentice-Hall do Brasil, Ltda., *Rio de Janeiro*

This text is dedicated to my Father, who taught me the practical application of knowledge plus problem solving by using our natural God given talent.

Len Mrachek

CONTENTS

Section One: **Arithmetic**

Chapter 1 FUNDAMENTAL OPERATIONS OF ARITHMETIC, 1
Chapter 2 COMMON FRACTIONS, 14
Chapter 3 DECIMALS, 35
Chapter 4 PERCENTAGES, 48
Chapter 5 POWERS AND ROOTS, 67

Section Two: **Basic Algebra**

Chapter 6 DEFINITIONS AND BASIC OPERATIONS OF ALGEBRA, 83
Chapter 7 SIMPLE EQUATIONS AND FORMULAS, 100
Chapter 8 FORMULA EVALUATION, 117
Chapter 9 FORMULA TRANSPOSITION, 130
Chapter 10 RATIO AND PROPORTION, 148

Section Three: **Applied Geometry**

Chapter 11 BASIC DEFINITIONS AND PROPERTIES OF GEOMETRY, 165
Chapter 12 PERIMETERS AND AREAS OF PLANE GEOMETRIC FIGURES, 179
Chapter 13 SURFACE AREAS AND VOLUMES OF GEOMETRIC FIGURES, 201
Chapter 14 CONSTRUCTION OF SIMPLE GEOMETRIC FIGURES, 223

Section Four: **Trigonometry**

Chapter 15 FUNDAMENTALS OF TRIGONOMETRY, 241
Chapter 16 SOLUTION OF RIGHT TRIANGLES, 258
Chapter 17 SOLUTION OF OBLIQUE TRIANGLES, 279

Section Five: **Metrication**

Chapter 18 FUNDAMENTALS AND APPLICATIONS OF THE METRIC SYSTEM, 295

BASIC TECHNICAL COLLEGE MATHEMATICS

section one: arithmetic

FUNDAMENTAL OPERATIONS OF ARITHMETIC

OBJECTIVES

1. To review the fundamental definitions and skills of arithmetic as a foundation for further mathematical studies.
2. To provide an opportunity to improve skills and speed in the fundamental operations.
3. To solve applied problems using the fundamental operations.

SELF-TEST

S–1–1.
```
  474
  439
   56
 +822
 ────
 1791
```

S–1–2.
```
 $1598
    39
 + 199
 ─────
 $1837
```

S–1–3.
```
  789
 -454
 ────
  335
```

S–1–4.
```
 $1239
 - 444
 ─────
   795
```

S–1–5.
```
 47,463
-16,587
 ──────
 30,876
```

S–1–6.
```
 $1289
 ×  12
 ─────
  2578
 1289
 ─────
 $15468
```

S–1–7.
```
   539
  ×321
  ────
   539
  1078
 1617
 ──────
 173,019
```

S–1–8.   7)287   =41

S–1–9.   4)$556   =$139
```
 15
 12
 36
```

S–1–10. 408,060 ÷ 20 =
```
         30403
    20)408060
        40
         80
```

S-1-11. An electrician worked 48, 42, 44, and 50 hours. Find the total number of hours he worked.

S-1-12. Find the total points scored in the following NBA games: 83, 79, 124, 146, and 116.

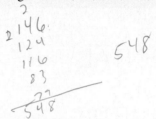

S-1-13. Kay Cobb's brother, Ruferthana, received checks for $586, $1429, $2246, $429, and $1842 as the deli manager in the supermarket. How much did he make?

$06532.

S-1-14. A painter used 28 gallons of paint from a 55-gallon drum. How much paint was left?

27 gallons

S-1-15. A water meter read 2412 cubic feet on September 1 and 2901 cubic feet on October 1. Find the number of cubic feet of water used during the month of September.

489 cubic feet.

S-1-16. A boat and trailer together weigh 1235 pounds. The trailer weighs 359 pounds. How much does the boat weigh?

876 pounds

S-1-17. Al Eisele's car odometer read 93,842 at the beginning of the summer and 99,326 at the end of the summer. How many miles did he drive during the summer?

5484 miles

S-1-18. If there are 5280 feet in 1 mile, how many feet are there in 12 miles?

63,360 miles

S-1-19. If gasoline costs 99 cents per gallon, find the cost of 125 gallons.

$123.75

S-1-20. A test tube weighs 14 grams. Find the weight of a gross of test tubes. (144 = 1 gross.)

2,016 grams

S-1-21. If a ham weighs 17 pounds, how many 1-ounce slices will you be able to get from this ham? (16 oz = 1 lb.)

S-1-22. How long would a steel bar need to be to make 18 pieces 17 inches long? (Disregard the amount of material lost in cutting.)

S-1-23. How much wire would be needed to wrap 14 coils if 2528 inches of wire is needed for each coil?

S-1-24. The "Front Four" of a football team are listed in the program as weighing 280, 265, 260, and 275 pounds. What is their average weight?

S-1-25. Find the value of D in Figure 1-1.

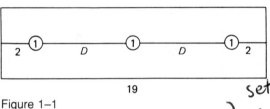

Figure 1-1

INTRODUCTION

Jack wanted to lower the cost of operating his car and decided to experiment as to which type of gasoline, unleaded or super unleaded, would provide the greater cost saving. He knew the price of super unleaded gasoline was generally about 10 cents per gallon more than the price of unleaded gasoline and that his car could use either gasoline. The choice of gasolines would be decided by the cost of gasoline per mile.

Jack knew that more research would be needed if he considered the total performance of his car and the efficiency of his engine. But for the moment, he was concerned with the cost of gasoline per mile. To do these tests accurately, Jack needed knowledge and skill in the use of arithmetic. What skills will help him solve this problem?

First, Jack must have an understanding of arithmetic. A major cause of failure in mathematics is a lack of knowledge of what to do. Second, Jack must not only know what to do, but must do the operations accurately. In the following sections you will learn the things Jack needed to solve his problems. It will benefit you to learn them.

DEFINITIONS

The numbers used in counting are called integers. An integer is a whole or natural number used in counting. Each integer is made up of one or more *digits*. Thus the number 8 has one digit, the number 23 has two digits (2 and 3), and the number 4146 has four digits (4, 1, 4, and 6). Each digit has a place value. An *even integer* is a number whose last digit (reading from left to right) is divisible by 2. Examples of even numbers are 2, 4, 6, 8, 12, and 96. An *odd integer* is a number whose last digit is not divisible by 2. Thus 3, 5, 7, 9, 17, and 361 are examples of odd numbers. A *prime number* is a number that can be divided only by itself and the number 1. The first eight prime numbers are: 2, 3, 5, 7, 11, 13, 17, and 19. The number 1 may or may not be considered a prime number.

Numbers that are combined or added to give a total or sum are called *addends*. Addition is indicated by the + symbol, called the *plus sign* of operation. The plus sign of operation tells you to add or combine the numbers to obtain a total. The result of combining the addends is called the *sum*.

$$\begin{array}{rl} 12 & \text{(addend)} \\ +25 & \text{(addend)} \\ \hline 37 & \text{(sum)} \end{array}$$

The purpose of *subtraction* is to find out how much larger one number is than another. The larger number is placed on top of the smaller number. The larger number is called the *minuend*, and the smaller number is called the *subtrahend*. The result is called the *difference*. The sign of operation of subtraction is the *minus sign* ($-$). Subtraction is the opposite operation of addition.

$$\begin{array}{rl} 25 & \text{(minuend)} \\ -12 & \text{(subtrahend)} \\ \hline 13 & \text{(difference)} \end{array}$$

You may wish to add the same number several times. For example, if you wanted to find out how many inches there are in 7 feet, you would add seven 12's together to arrive at the answer of 84.

$$\begin{array}{r} 12 \\ 12 \\ 12 \\ 12 \\ 12 \\ 12 \\ +12 \\ \hline 84 \end{array} \quad \text{or} \quad \begin{array}{rl} 12 & \text{(multiplicand)} \\ \times\ 7 & \text{(multiplier)} \\ \hline 84 & \text{(product)} \end{array}$$

You could multiply 12 by 7 and obtain the same answer of 84. This process is called *multiplication* and is considered a rapid form of addition. It saves time and decreased the chances of making mistakes. When two numbers are multiplied, the resulting answer is called the *product*. The top number is called the *multiplicand*, and the bottom number is called the *multiplier*. The sign of operation of multiplication is the *times sign* ($\times$).

You may wish to divide a quantity into two or more parts. For example, if you wanted to find out how many dozen rolls you would need for 72 people, you could keep on subtracting 12's from 72 until you reach a number that was less than 12. You would then count the number of 12's you had subtracted to find out how many dozen rolls you would need.

$$\begin{array}{rrrrrr} 72 & 60 & 48 & 36 & 24 & 12 \\ -12 & -12 & -12 & -12 & -12 & -12 \\ \hline 60 & 48 & 36 & 24 & 12 & 0 \end{array}$$

You could also solve this problem by dividing 72 by 12 to arrive at the same answer, 6.

$$\text{(divisor)}\ 12\overline{)72}\ \begin{array}{l}6\ \text{(quotient)}\\ \text{(dividend)}\end{array}$$

This process, called *division*, may be considered as rapid subtraction. When one number is divided by another, the resulting answer is called the *quotient*. The number that is used to divide is called the *divisor*, and the number that is divided is called the *dividend*. The sign of operation of division is the *divide sign* ($\div$). When this sign appears in a problem, it tells you to divide the second number into the first number. For example, $72 \div 12$ tells you to divide 72 by 12. When you attempt to solve the problem by division, you rewrite the problem so you can more easily arrive at the correct answer. Thus you could rewrite $72 \div 12$ as $12\overline{)72}$.

CONTENT

The four basic operations of mathematics are addition, subtraction, multiplication, and division. They are the essential tools for every technician and tradesperson. The following examples illustrate how to use the basic operations of mathematics.

Example 1–1:   Addition

PROBLEM: Jack drove 169 miles the first week, 96 miles the second week, 174 miles the third week, and 191 miles the fourth week, using super unleaded gasoline. How many miles did he drive using super unleaded gasoline?

SOLUTION: Add (from top to bottom).

$$\begin{array}{rl} 169 & \text{(addend)} \\ 96 & \text{(addend)} \\ 174 & \text{(addend)} \\ +191 & \text{(addend)} \\ \hline 630 & \text{(sum)} \end{array}$$

CHECK: Re-add from bottom to top.

Example 1–2:   Subtraction

PROBLEM: How many miles did Jack drive on a trip if the odometer read 79,386 at the start and 81,234 at the end of the trip?

SOLUTION: Subtract 79,386 from 81,234; borrow as necessary.

$$\begin{array}{rl} 81,234 & \text{(minuend)} \\ -79,386 & \text{(subtrahend)} \\ \hline 1,848 & \text{(difference)} \end{array}$$

CHECK: Add.

$$\begin{array}{rl} 79,386 & \text{(subtrahend)} \\ +\ 1,848 & \text{(difference)} \\ \hline 81,234 & \text{(minuend)} \end{array}$$

Example 1–3: Multiplication

PROBLEM: Jack paid 134 cents for each gallon of super unleaded gasoline. How much did he pay for the 18 gallons of super unleaded gasoline?

SOLUTION: Multiply the gallons by the price per gallon.

```
    134       price (cents) per gallon (multiplicand)
  × 18        gallons of unleaded (multiplier)
  ----
   1072
    134
  ----
   2412       total cost for
              unleaded gasoline (product)
```

CHECK: Divide 18 into 2412.

```
                134       (quotient)
  (divisor)  18)2412      (dividend)
                18
                --
                 61
                 54
                 --
                  72
                  72
```

Example 1–4: Division

PROBLEM: Jack used 21 gallons of unleaded gasoline to drive 630 miles. How many miles did he travel per gallon of gasoline?

SOLUTION: Divide the miles driven (630) by the total number of gallons of leaded gasoline (21).

```
                30 miles per gallon    (quotient)
  (divisor)  21)630 miles driven       (dividend)
                63
                --
                 0
```

CHECK: Multiply 30 times 21.

```
        30         (multiplicand)
      × 21         (multiplier)
      ----
        30
        60
      ----
       630         (product)
```

Example 1–5: Multiplication and Addition

PROBLEM: Jack filled his car with gasoline five times and it took 14 gallons to fill it the sixth time. If the capacity of the fuel tank is 23 gallons, how much gasoline has he purchased?

SOLUTION: Multiply 5 times 23 gallons and add 14 gallons.

```
    23 gallons         115 gallons
  ×  5               +  14
  ----               -----
   115 gallons         129 gallons
```

CHECK: Re-do the multiplication and then re-add the addition.

Example 1–6: Subtraction

PROBLEM: Jack wanted to find out which cost him less to use, unleaded or super unleaded gasoline. It cost Jack 2604 cents for unleaded gasoline to drive the 630 miles and 2412 cents for super unleaded to drive the same distance. When driving the 630 miles, how much less did it cost Jack for the super unleaded than for the unleaded?

SOLUTION: Subtract the cost of unleaded (2412 cents) from the cost of superunleaded (2604 cents).

```
    2604     cost (cents) for unleaded gasoline
             (minuend)
             cost (cents) for superunleaded gasoline
  − 2412     (subtrahend)
  ------
     192     less using unleaded gasoline (difference)
```

CHECK: Add 2412 and 192.

```
    2412     (addend)
  +  192     (addend)
  ------
    2604     (sum)
```

Example 1–7: Addition and Division

PROBLEM: Jack wondered how many miles he drove on average each week. The miles driven during the nine weeks were: 169, 96, 174, 191, 154, 117, 137, 211, and 128. Find the average miles driven per week.

SOLUTION: Add the total number of miles driven each week.

```
    169     (addend)
     96     (addend)
    174     (addend)
    191     (addend)
    154     (addend)
    117     (addend)
    137     (addend)
    211     (addend)
  + 128     (addend)
  -----
   1377
```

Then divide the sum by the total number of weeks.

```
                       153 average miles   (quotient)
(divisor)   9 weeks )1377 miles driven     (dividend)
                     9
                     ──
                     47
                     45
                     ──
                      27
                      27
                      ──
```

CHECK: First, re-add from bottom to top. Then multiply the average mileage by the number of weeks.

```
    153    (multiplicand)
  ×   9    (multiplier)
  ─────
   1377    (product)
```

EXERCISE 1-1

Do the following problems using the basic arithmetic operations. Be sure to check your results.

1-1. $9.98
        4.49
      + 3.29
      ───────
       $17.76

1-2.   10.8 gallons
        5.7 gallons
        9.4 gallons
      + 8.0 gallons
      ────────────
       24.9 gallons

1-3.   1673 pounds
       1841 pounds
       1397 pounds
     + 224 pounds
     ─────────────
       5135 pds

1-4.   386 miles
       513 miles
       121 miles
     + 86   miles
     ─────────────
       1106 miles

1-5.   4.25   hours
       8.75   hours
      10.25   hours
       4.25   hours
      ──────────────
       31.75 hrs

1-6. The air distance from Atlanta to Boston is about 837 miles. From Atlanta to New Orleans it is about 393 miles. How far is it from Boston to New Orleans via Atlanta?

1230 miles

1-7. A wholesaler sells 504 fishing reels to one dealer, 103 to another, 97 to another, and has 321 remaining. How many did he have at the beginning?

1025 Reels

1-8. The area in square miles of the following states is: Alaska, 586,400; Alabama, 54,609; and Arkansas, 53,104. What is the total area of these states?

694,113 square miles

1-9. Find the total area (in square miles) of the five largest countries in the world. They are: the former Soviet Union, 8,599,300; Canada, 3,851,809; China, 3,691,500; United States, 3,675,633; and Brazil, 3,286,478.

23,104,720 sq. mi

1–10. Dayton's service department did the following business for labor and parts in 1 hour:

	Labor	Parts
Hocker Hanson	$22.50	$19.50
Ducky Dunbar	22.75	27.00
Bird Benson	25.00	22.00
Phred Freeberg	20.50	19.85

(a) Find the total income from labor. *70.75*
(b) Find the total income from parts. *88.35*
(c) Find the total income from labor and parts for the 1 hour. *179.1*

1–11. If a video camera costs $1495 at one store and $999 at a second store, how much would you save buying the camera at the second store?

496 miles

1–12.   28.8 gallons
      − 8.9 gallons
      19.9 gallons

1–13.   17,001 miles
      − 9,314 miles
      7687 miles

1–14.   50,000 miles
      − 19,876 miles
      30124 miles

1–15. The fuel capacity of a jet plane is 35,000 pounds. After a flight, it takes 27,673 pounds of fuel to refuel the plane. How much fuel was used on the flight?

35,000
27,673
7,327

1–16. How much greater is 12,000 than 9249?

2,751

1–17. The fuel capacity of a 1986 Chevrolet van is listed at 36 gallons. If it takes 19 gallons of gasoline to fill the tank, how much fuel was left in the tank?

17 gallons

1–18. A doctor recommended a dosage of 450 cubic centimeters per day and then changed it to 750 cubic centimeters. How much did he increase the dosage?

300 cubic centimeters cc

1–19. The warranty on a new Chrysler car is 70,000 miles. The odometer shows 39,875 miles. How many miles are left on the warranty?

30,125 miles

1–20. A television set was priced at $579. Then it was marked down to $519. Later it was marked down further, to $495. What was the total discount?

$84.

1–21. 98
     × 8
     784

1–22. 2406
     × 39
     93,834

1–23. 5280
     × 707
     3,732,960

1–24. $1498
     × 9
     13,482

1–25. $3.89
     × 14
     54.46

1–26. Office space at a new office complex rents for $12 per square foot. What would the rent be for an office space of 1800 square feet?

21600 sqft

1–27. If a truck averages 54 miles per hour, how many miles will the truck cover in 112 hours?

6,048 miles

1–28. If it takes 3 hours of flying time between Seattle and Minneapolis, how much flying time would it take to make three round trips?

3×2=6×3 = 18 hrs

1–29. Tony Beldon bought three compact disks (CDs) at $15 each and two CDs at $18 each. How much did he spend for the CDs?

$81.00

1–30. What is the cost of 6 Saturns at $19,995 each?

119,970

1–31. 6)528
       88
       48
       48

1–32. 17)663
       39
       51
       150

1–33. $22.50 ÷ 6 =
       3.75

1-34. 5412 ÷ 451 =   12

1-35. 16)$223,968

   13998.

1-36. The total for a catering bill for a wedding was $7209. If there were 267 guests, what was the cost for each guest?

$27.00 per guest

1-37. The starting salary for a nurse at Med Center One is $23,340 per year. How much is this per month?

$1,945.00

1-38. If you traveled 496 miles on 16 gallons of gas, what was your average mileage per gallon?

31 mpg

1-39. A lot in a posh suburb sold for $169,884. If the lot contained 4356 square feet, what was the cost per square foot?

$.39

1-40. Tom Melchior spent $3588 for groceries for his family during the year. What was his average weekly cost for groceries? (There are 52 weeks in 1 year.)

$69. per week

1-41. What is the average of 100, 240, 440, and 880?

415.00

1-42. What is the average age of a family whose father is 52; mother, 48; daughter, 25; daughter, 24; son, 21; daughter, 19; and daughter, 14?

29 years

1-43. What is the average resistance in ohms of a group of resistors whose values are 570, 450, 630, and 710?

590 ohms

1-44. What is the average dosage per patient if the following dosages (in cubic centimeters) are prescribed: 420, 180, 100, 200, and 225?

225 cc

1-45. The weekly salaries at the supermarket are as follows: Velda, $304; Jake, $289; Thelma, $273; Phred, $310; Harold, $265; and Brucy, $293. What is the average *annual* salary? (52 weeks/year)

15,028.

Chapter 1 / Fundamental Operations of Arithmetic

1-46. Find the missing distance in Figure 1-2.

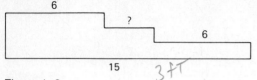

Figure 1-2

1-47. Find the missing dimension in Figure 1-3.

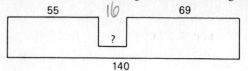

Figure 1-3

1-48. What is the thickness of the pipe shown in Figure 1-4?

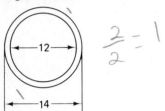

Figure 1-4

1-49. Find the missing dimension in Figure 1-5.

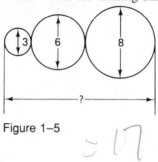

Figure 1-5

1-50. Find the missing dimension in Figure 1-6.

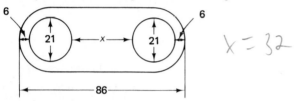

Figure 1-6

THINK TIME

As you worked the examples and exercises in this chapter, you will recall using the four basic operations (addition, subtraction, multiplication, and division). Think of the things you have learned in this chapter that will help you on the job. If you have an opportunity to discuss and review some of these with a teacher or a friend, it would be to your advantage. The more capable you are in basic skills, the more successful you will be in many things.

PROCEDURES TO REMEMBER

The basic operations of arithmetic are:

1. Reading the signs of operation correctly.
2. Performing the operations by the most efficient method and showing all necessary work.
3. Checking your work to make sure the answers are reasonable and correct. Methods of checking are as follows:
    (a) *Addition:* Re-add from bottom to top.
    (b) *Subtraction:* Add the difference to the subtrahend to obtain the minuend.
    (c) *Multiplication:* Divide the product by the multiplier to obtain the multiplicand.
    (d) *Division:* Multiply the quotient by the divisor to obtain the dividend.

CHAPTER SUMMARY

1. An *integer* is a whole or natural number used in counting.
2. An *even number* is a number that is divisible by 2.
3. An *odd number* is a number that is *not* divisible by 2.
4. A *prime number* is a number that can be divided only by itself and 1.
5. The numbers that are added to give a *sum* or *total* are called *addends*. The plus sign (+) indicates addition. This is one of the *signs of operation* of mathematics.
6. *Subtraction* is the arithmetic operation of finding the *difference* between the *minuend* (top number) and the *subtrahend* (bottom number). The sign of operation is the minus or negative sign (−).
7. The *product* is obtained when you multiply the *multiplicand* (top number) by the *multiplier* (bottom number). This is called *multiplication*. The sign of operation is ×.
8. The *quotient* is obtained when you divide one number, the *divisor*, into another number, the *dividend*. The sign of operation of division is ÷. Divide the second number into the first. Thus we may show $20 \div 4$ as $4\overline{)20}$.

CHAPTER TEST

Solve the following problems using the fundamental operations of arithmetic. Show all necessary work.

T–1–1. $1529
       + 164
       ──────
         1693

T–1–2. $18,398
       −   8,453
       ────────
           9,945

T–1–3. $3829
           546
            29
           498
       +   39
       ──────
          4941

T–1–4. $139
       ×   15
       ──────
          695
         1390
       ──────
         2085

T–1–5. 9)2016   = 224

T–1–6. If a master carpenter earns $41,288 in a year, what are his average weekly earnings?

794.

T–1–7. Lawn mowers for cutting grass on a golf course cost $3295 each. How much would three of them cost?

$9,885.

T–1–8. If a truck driver earns $170 a day, how much would he make in 250 working days?

$42,500

T–1–9. If water boils at 212°F at sea level and at 193°F in the McKinley Range of mountains, what is the difference in boiling points?

19°F

T–1–10. An electrician used the following lengths of number 14 wire: 63, 132, 86, 168, and 1024 inches. How much wire did he use?

1473' Inches

Chapter 1 / Fundamental Operations of Arithmetic

T–1–11.   8877
          2345
           618
          6824
        +   36

        18,700

T–1–12. $20,382
             527
             322
         + 1,296

          22,527

T–1–13.   23,273
        – 16,439

          6,834

T–1–14.   999
        × 99

        98901

T–1–15. 84
          $408\overline{)34{,}272}$

T–1–16. A washing machine in a bottling plant washes 3024 bottles in 12 hours. How many bottles does it wash per hour?

252 Bottles per hr.

T–1–17. A running back carried the ball 22 times in one game for a total yardage of 198. What was the average yardage he gained per carry?

9 yards per carry

T–1–18. A garment worker makes 368 golf shirts in a 8-hour day. How many shirts are made per hour?

$8\overline{)368}$ = 46

T–1–19. An electrical generator produces 2700 watts of power. How many 75-watt light bulbs will this light?

36 Bulbs

T–1–20. The shipping weight of each of 15 electronic testing scopes is 219 pounds, including the carton. What is the total shipping weight?

3285 LBS.

T–1–21. A farmer wants to paint his barn, which has 14,950 square feet. How many gallons of paint will be needed if each gallon covers 650 square feet?

23 gallons

T-1-22. Vinny Vangogo, a painter, buys 22 gallons of paint at $13 per gallon and 11 gallons at $17 per gallon. How much did the paint cost?

$473

T-1-23. Some duck hunters purchased a raft at $253 and a tent at $195. How many hunters were involved if each paid an equal amount of $112?

4 Hunters

T-1-24. Members of the Jog for Joy club were comparing the distance they jogged each month. The distances stated were: 193, 124, 86, 113, and 59 km. Find the average distance jogged per month.

115 Km.

T-1-25. Find T in Figure 1-7.

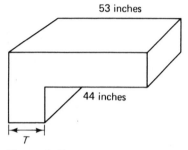

Figure 1-7

$T = 9$ inches

Chapter 1 / Fundamental Operations of Arithmetic

2

COMMON FRACTIONS

OBJECTIVES

1. To review the basic concepts and definitions of fractions.

2. To provide the opportunity to develop your skills in working with fractions.

SELF-TEST

This test will determine your need to develop your skills with fractions. If you are able to work these problems, move on to decimals. Show your work; this will indicate areas where you may need to improve.

S–2–1.
$$\frac{5}{6} \quad \frac{10}{12}$$
$$\frac{3}{4} \quad \frac{9}{12}$$
$$+\frac{1}{3} \quad \frac{4}{12}$$
$$\frac{23}{12} = 1\frac{11}{12}$$

S–2–2.
$$24\frac{1}{3} \quad \frac{2}{6}$$
$$+10\frac{5}{6}$$
$$34\frac{7}{6} = 35\frac{1}{6}$$

S–2–3.
$$6\frac{3}{4} \quad \frac{9}{12}$$
$$7\frac{1}{2} \quad \frac{6}{12}$$
$$+2\frac{2}{3} \quad \frac{8}{12}$$
$$15 \quad \frac{23}{12} = 16\frac{11}{12}$$

S–2–4. $\frac{4}{5} - \frac{1}{3} =$

$$\frac{12}{15} - \frac{5}{15} = \frac{7}{15}$$

S–2–5.
$$15\frac{2}{5} \quad \frac{6}{15} \quad \overset{14}{} \overset{21}{}$$
$$-5\frac{2}{3} \quad \frac{10}{15}$$
$$9 \quad \frac{11}{15}$$

S–2–6.
$$\overset{33}{34}\frac{1}{3} \quad \frac{4}{12} = \frac{16}{} \overset{8}{} \frac{\cancel{x}}{6}$$
$$-29\frac{5}{6} \quad \frac{10}{12} \quad \frac{5}{6}$$
$$4 \quad \frac{4}{12} = 4\frac{8}{}$$
$$= 4\frac{1}{2}$$

S–2–7. $\overset{1}{\cancel{\frac{2}{3}}} \times \overset{2}{\underset{9}{\cancel{\frac{6}{18}}}} = \frac{4}{18} = \frac{2}{9}$

S–2–8. $3\frac{1}{2} \times 1\frac{3}{4} =$

$$\frac{7}{2} \times \frac{7}{4} = \frac{49}{8} = 6\frac{1}{8}$$

S-2-9. $24\frac{1}{2} \div 18\frac{5}{6} =$

$\frac{49}{2} \times \frac{6}{10} = \frac{147}{113}$

S-2-10. $12\frac{1}{2} \div 18\frac{5}{6} = \quad 75/113$

S-2-11. $\frac{3}{4} \div 12 = \frac{3}{4} \times \frac{1}{12} = \frac{1}{16}$

S-2-12. $\frac{7}{8} \times \frac{7}{16} \times \frac{21}{32} = \frac{14}{24} = \frac{7}{12}$

S-2-13. $12\frac{1}{2} \div 4\frac{2}{3} \div 8\frac{3}{8} = \frac{150}{469}$

S-2-14. $3\frac{2}{3} \times 1\frac{3}{5} \times 2\frac{1}{12} = \frac{110}{9} \quad 12\frac{2}{9}$

S-2-15. $\dfrac{\frac{7}{8} - \frac{1}{2}}{4} = \frac{3}{32}$

S-2-16. $\dfrac{3\frac{1}{2} + \frac{3}{4}}{3\frac{1}{2} \times 2\frac{1}{4}} = \frac{34}{63}$

S-2-17. $\dfrac{1\frac{1}{2} - \frac{3}{4}}{1\frac{5}{6} \div 6} = 2\frac{5}{11}$

S-2-18. Find the length A in Figure 2–1.

$A = \frac{17}{10} - \left(\frac{3}{10} + \frac{3}{10} + \frac{5}{2}\right) =$

$A = \frac{17}{10} - \frac{11}{10} = A = \frac{6}{10} = \boxed{\frac{3}{5}}$

S-2-19. Find length B in Figure 2–1.

$2\left(1\frac{7}{10}\right) - \frac{1}{2}$

$2 \cdot \frac{17}{10} - \frac{1}{2}$

$\frac{17}{5} - \frac{1}{2}$

$\frac{34}{10} - \frac{5}{10} = \frac{29}{10} = 2\frac{9}{10}$

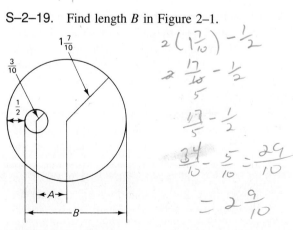

Figure 2–1

S-2-20. An oil stock sold for $11\frac{1}{2}$ a share when the stock market opened. The stock closed at $12\frac{1}{8}$. How much did the stock gain?

$12\frac{1}{8} - 11\frac{1}{2} = 12\frac{1}{8} - 11\frac{4}{8}$

$11\frac{9}{8}$
$-11\frac{4}{8}$
$\frac{5}{8}$

Chapter 2 / Common Fractions

S-2-21. Steak sells for $3 per pound. What is the total cost for steaks weighing $1\frac{1}{16}$, $1\frac{1}{4}$, $1\frac{3}{8}$, and $1\frac{5}{16}$ pounds?

S-2-22. Total resistance is the sum of the individual resistances. What is the resistance of three motors connected in a series if their resistances are $1\frac{1}{25}$, $2\frac{3}{50}$, and $5\frac{37}{150}$?

S-2-23. The gas tank of a car shows $\frac{3}{4}$ full. If the full tank contains 24 gallons and gasoline costs $2 per gallon, how much is the gas worth that is in the tank?

S-2-24. A truck travels $12\frac{4}{10}$ miles on a round trip from a plant to a construction site. On the eighth trip, the truck stops at the construction site. How many miles has the truck been driven?

S-2-25. A master painter can paint a garage in 6 hours. An apprentice can paint the same garage in 9 hours. Working together, how much of the garage can they paint in 1 hour?

INTRODUCTION

Common fractions are an important part of our day-to-day living. We use fractions to indicate time—"I will meet you in a half hour"; quantity—"You have only a quarter of a tank of gasoline left"; distance—"The horses have passed the three-quarter pole." Fractions are a part of the language of industry and technology. Terms such as "$\frac{3}{32}$-inch welding rods," "$\frac{5}{8}$-inch plywood," "three-quarters of a cup of flour," "$\frac{9}{64}$-inch drill," "three-quarter-ton truck," and "one-third off" are used many times. Therefore, if we are to survive and contribute to society, it is essential for us to have fraction skills.

DEFINITIONS

A *fraction* is the division of one number by another number. Examples of fractions are $\frac{1}{8}, \frac{2}{3}$, and $\frac{6}{7}$. The number above the line is called the *numerator* and the number below the line is called the *denominator*. The numerator and denominator are the terms for a fraction. Thinking of the great state of North Dakota—ND—will help you remember that the numerator precedes (is above) the denominator. A fraction whose denominator is larger than its numerator is called a *proper fraction*. Examples of proper fractions are $\frac{3}{4}, \frac{7}{8}$, and $\frac{3}{16}$.

A fraction whose numerator is greater than its denominator is called an *improper fraction*. Examples of improper fractions are $\frac{5}{4}, \frac{32}{9}$, and $\frac{8}{5}$.

A whole number and a fraction written together are called *mixed numbers*. Examples of mixed numbers are $4\frac{3}{4}, 5\frac{3}{16}$, and $6\frac{7}{8}$.

A fraction that has a fraction or fractions in the numerator and/or denominator is called a *complex fraction*. Examples of complex fractions are

$$\frac{\frac{6}{7}}{\frac{3}{5}}, \quad \frac{\frac{2}{3}}{\frac{1}{2}}, \quad \text{and} \quad \frac{\frac{3}{1}}{\frac{7}{8}}$$

To *reduce* a fraction to lowest terms, divide the numerator and the denominator by the same number.

A competent technician or skilled craftsperson needs to perform operations with fractions. These operations are addition, subtraction, multiplication, and division. We start with the multiplication of fractions, which is the easiest operation.

MULTIPLICATION OF FRACTIONS

Example 2–1

PROBLEM: $\dfrac{2}{3} \times \dfrac{1}{5} =$

SOLUTION: Multiply the numerator (2) by the numerator (1) and the denominator (3) by the denominator (5).

$$\frac{2}{3} \times \frac{1}{5} = \frac{2 \times 1}{3 \times 5} = \frac{2}{15}$$

Example 2–2

PROBLEM: $\dfrac{3}{4} \times \dfrac{7}{15} =$

SOLUTION: Before multiplying fractions, reduce the fractions to the lowest terms. Reducing before multiplying will make the multiplication process easier. The reducing may be done by different methods.

METHOD 1: Divide the numerator (3) and the denominator (15) by 3. The number to be divided evenly into the numerator and the denominator is done by observation. Practice will refine your selection ability.

$$\frac{\overset{1}{\cancel{3}}}{4} \times \frac{7}{\underset{5}{\cancel{15}}} = \frac{1 \times 7}{4 \times 5} = \frac{7}{20}$$

METHOD 2: Divide the top (numerator) of one fraction (3) into the bottom (denominator) of another fraction (15). Thus the numerator (3) divides into the denominator (15) five times and into itself (3) once. A simple rule to remember is "*top* into *bottom*" and "*bottom* into *top*." After reducing the fractions to lowest terms, multiply the numerators 1 and 7 and the denominators 4 and 5.

$$\frac{\overset{1}{\cancel{3}}}{4} \times \frac{7}{\underset{5}{\cancel{15}}} = \frac{1 \times 7}{4 \times 5} = \frac{7}{20}$$

Example 2–3

PROBLEM: $\dfrac{39}{9} \times \dfrac{45}{13} \times \dfrac{1}{10} =$

SOLUTION: When reducing fractions to lowest terms, there may be several possible reductions. Make certain you have made all possible reductions before you multiply the fractions.

METHOD 1: Divide the numerator (39) and the denominator (13) by 13. Divide the denominator (9) and the numerator (45) by 9. Then divide the new numerator (4) and the denominator (10) by 5.

$$\frac{\overset{3}{\cancel{39}}}{\underset{1}{\cancel{9}}} \times \frac{\overset{\overset{1}{\cancel{5}}}{\cancel{45}}}{\underset{1}{\cancel{13}}} \times \frac{1}{\underset{2}{\cancel{10}}} = \frac{3 \times 1 \times 1}{1 \times 1 \times 2} = \frac{3}{2}$$

METHOD 2: Divide the *bottom* (denominator) (9) into the *top* (numerator) (45) and into 9. Divide the *bottom* (denominator) (13) into the *top* (numerator) (39) and into itself (13). Divide the new *top* (numerator) (5) into the *bottom* (denominator) (10) and into itself (5). After reducing fractions, multiply the numerators (3, 1, 1) together and the denominator (1, 2, 2) together.

$$\frac{\overset{3}{\cancel{39}}}{\underset{1}{\cancel{9}}} \times \frac{\overset{\overset{1}{\cancel{5}}}{\cancel{45}}}{\underset{1}{\cancel{13}}} \times \frac{1}{\underset{2}{\cancel{10}}} = \frac{3 \times 1 \times 1}{1 \times 1 \times 2} = \frac{3}{2}$$

Then convert the improper fraction to a mixed number. Divide the numerator (3) by the denominator (2). The remainder (1) is the numerator.

$$\frac{3}{2} = 1\frac{1}{2}$$

DIVISION OF FRACTIONS

The procedure for dividing fractions is similar to the procedure for multiplying fractions, with one major change. The first step in dividing fractions is to invert the second fraction or the fraction on the right. To *invert* means to interchange or reverse the positions of the denominator and the numerator. Thus the second fraction is literally turned upside down. Then proceed as in multiplication of fractions.

Example 2–4

PROBLEM: $\dfrac{4}{5} \div \dfrac{1}{3} =$

SOLUTION: After inverting the second fraction, reduce the fractions to lowest terms, multiply and convert to a mixed number. Remember, multiplication is done *after* the inverting, not before.

$$\dfrac{4}{5} \div \dfrac{1}{3} = \dfrac{4}{5} \times \dfrac{3}{1} = \dfrac{12}{5} = 2\dfrac{2}{5}$$

Example 2–5

PROBLEM: $\dfrac{5}{8} \div \dfrac{3}{16} =$

SOLUTION: After inverting the second fraction, reduce and multiply, then convert to a mixed number. Remember, multiplication is done *after* the inverting, not before.

$$\dfrac{5}{8} \div \dfrac{3}{16} = \dfrac{5}{\underset{1}{8}} \times \dfrac{\overset{2}{\cancel{16}}}{3} = \dfrac{10}{3} = 3\dfrac{1}{3}$$

Example 2–6

PROBLEM: $14 \div \dfrac{7}{8} =$

SOLUTION: Note there is no denominator for 14. When a number is shown without a denominator, it is understood the denominator is 1. After inverting the fraction on the right, reduce to lowest terms and multiply.

$$14 \div \dfrac{7}{8} = \dfrac{\overset{2}{\cancel{14}}}{1} \times \dfrac{8}{\underset{1}{\cancel{7}}} = \dfrac{2 \times 8}{1 \times 1} = \dfrac{16}{1} = 16$$

Example 2–7

PROBLEM: If half a pizza is to be divided equally among four people, how much will each person receive?

SOLUTION:

$$\dfrac{1}{2} \text{ pizza} \div 4 \text{ people} = \dfrac{1}{2} \div \dfrac{4}{1} = \dfrac{1}{2} \times \dfrac{1}{4} = \dfrac{1}{8}$$

Note each person will receive $\dfrac{1}{8}$ of the *whole* pizza. See Figure 2–2.

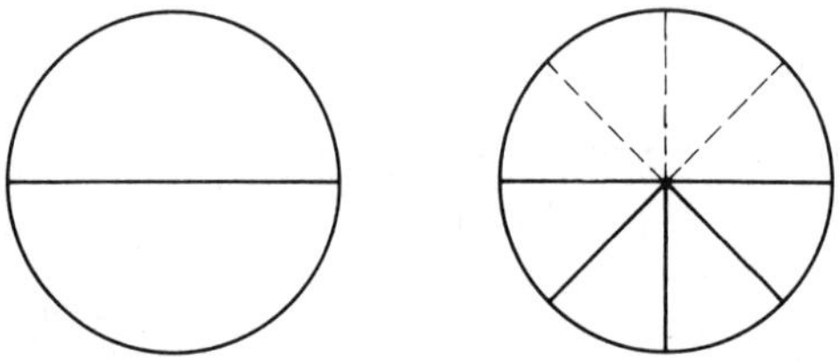

Figure 2–2

EXERCISE 2–1

Do the following problems to practice multiplying and dividing fractions.

2–1. $\dfrac{5}{9} \times \dfrac{2}{7} = \dfrac{10}{63}$

2–2. $\dfrac{3}{4} \times \dfrac{7}{8} = \dfrac{21}{32}$

2–3. $\dfrac{\overset{2}{\cancel{4}}}{1} \times \dfrac{3}{\underset{1}{\cancel{2}}} = 6$

2–4. $\dfrac{4}{7} \times \dfrac{2}{3} = \dfrac{8}{21}$

2–5. $\dfrac{3}{4} \times \dfrac{1}{2} = \dfrac{3}{8}$

2–6. $\dfrac{5}{\underset{1}{\cancel{6}}} \times \dfrac{\overset{2}{\cancel{12}}}{11} = \dfrac{10}{11}$

2–7. $\frac{3}{4} \times \frac{8}{7} \times \frac{14}{21} =$ 2–8. $\frac{3}{4} \times \frac{5}{12} \times \frac{7}{8} =$ 2–9. $\frac{3}{4} \div \frac{2}{3} =$

2–10. $\frac{1}{2} \div \frac{1}{3} =$ 2–11. $6 \div \frac{2}{3} =$ 2–12. $\frac{1}{10} \div \frac{3}{5} =$

2–13. $\frac{16}{5} \div \frac{18}{5} =$ 2–14. $\frac{4}{3} \div \frac{1}{3} =$ 2–15. $13 \div \frac{169}{15} =$

ADDITION OF FRACTIONS

Example 2–8

PROBLEM: $\frac{1}{2} + \frac{1}{3} =$

SOLUTION: To be able to add fractions, each fraction must have the same denominator. Thus the first step when adding fractions is to find the lowest common denominator. This means finding a number that is a multiple of each denominator. In this problem, find a number that is a multiple of 2 and 3. One way is to multiply the denominators together. However, this method does not always give the *lowest* common denominator. In our problem we can multiply the two denominators (2 × 3) and arrive at the common denominator of 6. Once a common denominator has been found, it is important to check if it is the lowest common denominator. In other words, is there a number less than 6 which is a multiple of 2 and 3? There is none, so 6 is the lowest common denominator for this problem.

The next step is to multiply each fraction by the lowest common denominator. Do this by using a fraction that is equal to 1. In this problem $\frac{6}{6}$ will be used. Note the outline of the number 1, representing 1 or unity, is superimposed over the fraction for illustration. This is only used to emphasize the importance of making certain that the fraction is equal to 1.

$$\frac{1}{2} \times \frac{6}{6} = \frac{3}{6}$$

$$+ \frac{1}{3} \times \frac{6}{6} = \frac{2}{6}$$

$$\frac{3}{6} + \frac{2}{6} = \frac{5}{6}$$

Then add the numerators. Note that *only* the numerators are added and the denominators remain the same.

Example 2–9

PROBLEM:
$$\frac{2}{5}$$
$$\frac{5}{8}$$
$$+ \frac{3}{10}$$

SOLUTION: Find the lowest common denominator for the denominators 5, 8, and 10. By multiplying 5 and 8, we obtain a possible common denominator; it must be a multiple of 10. Dividing 40 by 10, we obtain 4.

Chapter 2 / Common Fractions

Thus 40 is a common denominator because 5, 8, and 10 will all divide evenly into 40. Therefore, 40 is the lowest common denominator.

Then multiply each fraction by 1 or $\boxed{\dfrac{40}{40}}$:

$$\dfrac{2}{\underset{1}{\cancel{5}}} \times \boxed{\dfrac{\overset{8}{\cancel{40}}}{40}} = \dfrac{16}{40}$$

$$\dfrac{5}{\underset{1}{\cancel{8}}} \times \boxed{\dfrac{\overset{5}{\cancel{40}}}{40}} = \dfrac{25}{40}$$

$$+\dfrac{3}{\underset{1}{\cancel{10}}} \times \boxed{\dfrac{\overset{4}{\cancel{40}}}{40}} = \dfrac{12}{40}$$

An *alternative* second step is to multiply each fraction by a "special 1" that will give a common denominator of 40. To do this, consider the first fraction, $\dfrac{2}{5}$. Ask yourself "5 times what number equals 40?". With the answer 8, we can now multiply by 1 or $\boxed{\dfrac{8}{8}}$. Following the same steps for the fraction $\dfrac{5}{8}$, we obtain 1 or $\boxed{\dfrac{5}{5}}$ and $\boxed{\dfrac{4}{4}}$ for $\dfrac{3}{10}$. This method eliminates cancellation.

$$\dfrac{2}{5} \times \boxed{\dfrac{8}{8}} = \dfrac{16}{40}$$

$$\dfrac{5}{8} \times \boxed{\dfrac{5}{5}} = \dfrac{25}{40}$$

$$+\dfrac{3}{10} \times \boxed{\dfrac{4}{4}} = \dfrac{12}{40}$$

Then add the fractions. Remember, add only the numerators and write the sum over the common denominator, 40.

$$\begin{array}{r} \dfrac{16}{40} \\ \dfrac{25}{40} \\ +\dfrac{12}{40} \\ \hline \dfrac{53}{40} \end{array}$$

Convert the improper fraction $\dfrac{53}{40}$ to a mixed fraction by dividing the denominator (40) into the numerator (53) and writing the remainder as a fraction.

$$\begin{array}{r} 1\dfrac{13}{40} \\ 40\overline{)53} \\ \underline{40} \\ 13 \end{array}$$

Example 2–10

PROBLEM:

$$6\dfrac{1}{9}$$
$$1\dfrac{9}{14}$$
$$+4\dfrac{5}{21}$$

SOLUTION: Find the lowest common denominator for 9, 14, and 21. This could be done by multiplying 9 × 14 × 21. The result (2646) could be reduced to a lowest common denominator through trial-and-error division. However, this procedure is time consuming. Another way to find the lowest common denominator is to identify the *prime factors* or *prime numbers* of each denominator. Prime factors are numbers that when multiplied together will be equal to each denominator. In this problem, the prime factors of the denominator are

$$9 = 3 \times 3$$
$$14 = 2 \times 7$$
$$21 = 3 \times 7$$

The next step is to select those prime factors that will determine the lowest common denominator. Select the prime factors of the first fraction, 3 × 3, and then select additional prime factors that are different from 3 × 3. Use these additional prime factors only once. Thus the prime factors 3 × 3 × 7 × 2 are selected and the lowest common denominator is 126.

Then multiply each fraction by a "special 1" number. Method 1 uses $\dfrac{126}{126}$. Method 2 eliminates cancellation by selecting a number "1" that will produce the common denominator 126 through multiplication. Study both examples. To prevent mistakes, show all your work.

Section One / Arithmetic

Method 1 *Method 2*

$$6\frac{1}{9} \times \boxed{\frac{14}{126}} = 6\frac{14}{126} \qquad 6\frac{1}{9} \times \boxed{\frac{14}{14}} = 6\frac{14}{126}$$

$$1\frac{9}{14} \times \boxed{\frac{9}{126}} = 1\frac{81}{126} \qquad 1\frac{9}{14} \times \boxed{\frac{9}{9}} = 1\frac{81}{126}$$

$$+4\frac{5}{21} \times \boxed{\frac{6}{126}} = 4\frac{30}{126} \qquad +4\frac{5}{21} \times \boxed{\frac{6}{6}} = 4\frac{30}{126}$$

Then adding the whole numbers and the denominators.

$$\begin{array}{r} 6\frac{14}{126} \\ 1\frac{81}{126} \\ +\ 4\frac{30}{126} \\ \hline 11\frac{125}{126} \end{array}$$

SUBTRACTION OF FRACTIONS

Example 2–11

PROBLEM:

$$\begin{array}{r} 12\frac{7}{8} \\ -\ 4\frac{3}{16} \\ \hline \end{array}$$

SOLUTION: First, find the lowest common denominator for 8 and 16. By observation, this is easily determined to be 16. Then multiply each fraction by 1 or a "special 1."

Method 1 *Method 2*

$$12\frac{7}{8} \times \boxed{\frac{2}{16}} = 12\frac{14}{16} \qquad 12\frac{7}{8} \times \boxed{\frac{2}{2}} = 12\frac{14}{16}$$

$$-\ 4\frac{3}{16} \times \boxed{\frac{1}{16}} = 4\frac{3}{16} \qquad -\ 4\frac{3}{16} \times \boxed{\frac{1}{1}} = 4\frac{3}{16}$$

Subtract the denominators and whole numbers. Remember to subtract the subtrahend (bottom) numerator from the minuend (top) numerator. Denominators are *not* subtracted.

$$\begin{array}{r} 12\frac{7}{8} = \frac{14}{16} \\ -\ 4\frac{3}{16} = \frac{3}{16} \\ \hline 8\frac{11}{16} \end{array}$$

Example 2–12

PROBLEM:

$$\begin{array}{r} 24\frac{1}{6} \\ -\ 16\frac{5}{12} \\ \hline \end{array}$$

SOLUTION: Find the lowest common denominator, for 6 and 12, by observation; it is 12. Then multiply by 1 or a "special 1" as shown.

Method 1 *Method 2*

$$24\frac{1}{6} \times \boxed{\frac{2}{12}} = 24\frac{2}{12} \qquad 24\frac{1}{6} \times \boxed{\frac{2}{2}} = 24\frac{2}{12}$$

$$-16\frac{5}{12} \times \boxed{\frac{12}{12}} = -16\frac{5}{12} \qquad -16\frac{5}{12} \times \boxed{\frac{1}{1}} = -16\frac{5}{12}$$

However, you cannot subtract 5 from 2; therefore, we must borrow "1", or $\frac{12}{12}$, from 24. Add $\frac{12}{12} + \frac{2}{12} = \frac{14}{12}$, then subtract the numerator 5 from numerator 14 and subtract the whole numbers 16 from 23. Reduce $\frac{9}{12}$ by dividing the numerator and denominator by 3.

$$\begin{array}{r} \overset{23}{24}\frac{2}{12} + \frac{12}{12} = \frac{14}{12} \\ -16\frac{5}{12} \phantom{+ \frac{12}{12}} = \frac{5}{12} \\ \hline 7\frac{9}{12} \end{array}$$

Note: When "borrowing," rather than adding $\frac{12}{12}$ to the top fraction, merely add the numerator and denominator and place the sum above the numerator.

$$\frac{2}{12} \text{ is } 2 + 12 = 14.$$

Thus $7\frac{9}{12} = 7\frac{3}{4}$.

Chapter 2 / Common Fractions

EXERCISE 2-2

Solve the following addition and subtraction of fractions. Remember, you must have a common denominator when adding or subtracting fractions.

2-16. $\dfrac{2}{7}$
$+\dfrac{3}{7}$

Answer: $\dfrac{5}{7}$

2-17. $\dfrac{3}{5}$ — $\dfrac{6}{10}$
$+\dfrac{4}{10}$

Answer: 1

2-18. $\dfrac{5}{20}$
$+\dfrac{3}{5}$ — $\dfrac{12}{20}$

Answer: $\dfrac{17}{20}$

2-19. $\dfrac{5}{9}$ — $\dfrac{10}{18}$
$+\dfrac{3}{18}$

Answer: $\dfrac{13}{18}$

2-20. $1\dfrac{1}{2}$
$+1\dfrac{1}{8}$

Answer: $2\dfrac{5}{8}$

2-21. $\dfrac{1}{3}$ — $\dfrac{4}{12}$
$+\dfrac{1}{4}$ — $\dfrac{3}{12}$

Answer: $\dfrac{7}{12}$

2-22. $7\dfrac{1}{2}$
$+8$

Answer: $15\dfrac{1}{2}$

2-23. $\dfrac{11}{16}$
$+\dfrac{1}{2}$ — $\dfrac{8}{16}$

Answer: $\dfrac{19}{16}$, $1\dfrac{3}{16}$

2-24. $\dfrac{2}{3}$
$-\dfrac{1}{3}$

Answer: $\dfrac{1}{3}$

2-25. $\dfrac{2}{3}$ — $\dfrac{4}{6}$
$-\dfrac{1}{6}$

Answer: $\dfrac{3}{6}$, $\dfrac{1}{2}$

2-26. $3\dfrac{3}{4}$
$-2\dfrac{1}{2}$ — $\dfrac{2}{4}$

Answer: $1\dfrac{1}{4}$

2-27. $16\dfrac{8}{8}$ (15)
$-12\dfrac{5}{8}$

Answer: $3\dfrac{3}{8}$

2–28. $15\frac{31}{32}$
         $-\ 9$
         _____
         $6\frac{31}{32}$

2–29. $\frac{3}{8}$ $\frac{24}{64}$
        $\frac{13}{32}$ $\frac{26}{64}$
        $+\frac{5}{64}$ $\frac{5}{64}$
        _____
        $\frac{55}{64}$

2–30. $\frac{3}{20}$ $\frac{90}{60}$ $\frac{9}{60}$
        $\frac{2}{3}$ $\frac{40}{60}$
        $+7\frac{11}{12}$ $\frac{55}{60}$
        _____
        $\frac{104}{60}$ $1\frac{44}{60} = 1\frac{11}{15}$

SIMPLIFYING COMPLEX FRACTIONS

So far, you have learned how to change fractions from one form to another; perform the fundamental operations of multiplication, division, addition, and subtraction; and simplify proper and improper fractions by reducing them to lowest terms. Simplifying *complex fractions* will require all of these skills.

Example 2–13

   PROBLEM: Simplify $\dfrac{\frac{3}{4}}{\frac{1}{3}}$.

   SOLUTION: The line between the two fractions indicates division; thus rewrite the fractions in regular division form and solve.

$$\dfrac{\frac{3}{4}}{\frac{1}{3}} = \frac{3}{4} \div \frac{1}{3} = \frac{3}{4} \times \frac{3}{1} = \frac{9}{4} = 2\frac{1}{4}$$

Example 2–14

   PROBLEM: $\dfrac{1}{\frac{1}{2} + \frac{1}{3} + \frac{1}{5}} =$

   SOLUTION: This complex fraction contains three fractions in the denominator. When simplifying complex fractions, first perform the operations indicated in the numerator and denominator. The lowest common denominator for the fractions in the denominator is 30. Thus a "special 1," $\frac{30}{30}$, is used to determine the equivalent for each fraction.

$$\dfrac{1}{\frac{1}{2} + \frac{1}{3} + \frac{1}{5}} = \dfrac{1}{\underset{1}{\frac{1}{2} \times \frac{\boxed{30}}{30}} + \underset{1}{\frac{1}{3} \times \frac{\boxed{30}}{30}} + \underset{1}{\frac{1}{5} \times \frac{\boxed{30}}{30}}}$$

$$= \dfrac{1}{\frac{15}{30} + \frac{10}{30} + \frac{6}{30}}$$

Then add the fractions in the denominator.

$$\dfrac{1}{\frac{15}{30} + \frac{10}{30} + \frac{6}{30}} = \dfrac{1}{\frac{31}{30}}$$

Simplify by writing the fraction in normal division of fraction terms and divide.

$$\dfrac{1}{\frac{31}{30}} = \frac{1}{1} \div \frac{31}{30} = \frac{1}{1} \times \frac{30}{31} = \frac{30}{31}$$

EXERCISE 2–3

Simplify the following complex fractions. Remember to use all the skills you have learned to simplify the complex fractions.

2–31. $\dfrac{\frac{3}{16}}{\frac{1}{4}} = \frac{3}{16} \times \frac{4}{1} = \frac{3}{4}$

2–32. $\dfrac{\frac{1}{2}}{\frac{7}{8}} = \frac{1}{2} \times \frac{8}{7} = \frac{4}{7}$

2–33. $\dfrac{\frac{5}{3}}{\frac{3}{8}} = \frac{5}{1} \times \frac{8}{3} = \frac{40}{3} = 13\frac{1}{3}$

2–34. $\dfrac{\frac{7}{8}}{28} =$

2–35. $\dfrac{1\frac{1}{16}}{\frac{3}{16}} =$

2–36. $\dfrac{\frac{1}{7}}{3\frac{1}{7}} =$

2–37. $\dfrac{\frac{1}{2} + \frac{5}{8}}{\frac{5}{8} - \frac{1}{2}} =$

2–38. $\dfrac{\frac{5}{7} + \frac{3}{2}}{\frac{5}{3} \times 4} =$

2–39. $\dfrac{12\frac{1}{2} \times 4}{\frac{3}{4} \div \frac{6}{16}} =$

2–40. $\dfrac{\frac{23}{8} \div \frac{1}{8}}{69 \times \frac{1}{5}} =$

2–41. $\dfrac{8\frac{1}{3} \times 2\frac{1}{2}}{8\frac{1}{3} \div 2\frac{1}{2}} =$

2–42. $\dfrac{1}{\frac{1}{2} + \frac{3}{4} + \frac{5}{8} + \frac{7}{16}} =$

2–43. $\dfrac{5 \div \frac{1}{10}}{\frac{1}{2} + \frac{1}{5} + \frac{1}{10}} =$

2–44. $\dfrac{\frac{9}{6} \times \frac{2}{11}}{14\frac{3}{8} - 6\frac{17}{24}} =$

BLUEPRINT AND SPECIFICATION FRACTIONS

Example 2–15

PROBLEM: Find *A* and *B* given the blueprint shown in Figure 2–3.

SOLUTION: Find *A*. *A* is equal to the sum of all

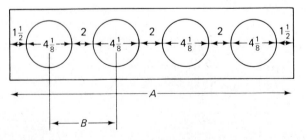

Figure 2–3

the distances. Thus all the fractions should be added together.

$$A = 1\frac{1}{2} + 4\frac{1}{8} + 2 + 4\frac{1}{8} + 2$$
$$+ 4\frac{1}{8} + 2 + 4\frac{1}{8} + 1\frac{1}{2}$$

METHOD 1: Find the lowest common denominator. By observation, it is 8. Multiply each fraction requiring change to the lowest common denominator (8) or $\frac{8}{8}$. Then add the fractions and reduce to lowest terms.

$$1\frac{1}{2} \times \frac{\cancel{8}^4}{\cancel{8}_1} = 1\frac{4}{8}$$

$$\begin{array}{rl} 4\frac{1}{8} & = 4\frac{1}{8} \\ 2 & = 2 \\ 4\frac{1}{8} & = 4\frac{1}{8} \\ 2 & = 2 \\ 4\frac{1}{8} & = 4\frac{1}{8} \\ 2 & = 2 \\ 4\frac{1}{8} & = 4\frac{1}{8} \\ + 1\frac{1}{2} \times \frac{\cancel{8}^4}{\cancel{8}_1} & = 1\frac{4}{8} \end{array}$$

METHOD 2: Another way to find A is to combine like dimensions by multiplying. Reduce to lowest terms and then add.

$$2 \times 1\frac{1}{2} = \frac{\cancel{2}^1}{1} \times \frac{3}{\cancel{2}_1} = 3$$

$$4 \times 4\frac{1}{8} = \frac{\cancel{4}^1}{1} \times \frac{33}{\cancel{8}_2} = \frac{33}{2} = 16\frac{1}{2}$$

$$+ 3 \times 2 = 6$$
$$A = 25\frac{1}{2}$$

PROBLEM: Find B. B is a center-to-center length. B is equal to $\frac{1}{2}$ of $4\frac{1}{8} + 2 + \frac{1}{2}$ of $4\frac{1}{8}$. Remember that "of" means multiply. So

$$B = \frac{1}{2} \times 4\frac{1}{8} + 2 + \frac{1}{2} \times 4\frac{1}{8}$$

Multiply by converting the mixed numbers, then add and reduce to lowest terms.

$$B = \frac{1}{2} \times \frac{33}{8} + 2 + \frac{1}{2} \times \frac{33}{8}$$
$$= \frac{33}{16} + 2 + \frac{33}{16}$$
$$= \frac{33}{16} + \frac{32}{16} + \frac{33}{16}$$

Thus $B = \frac{98}{16}$ or $6\frac{2}{16} = 6\frac{1}{8}$.

EXERCISE 2-4

Solve the following blueprint-reading problems.

2-45. Find A given the flat washer shown in Figure 2-4.

2-46. Find B given the flat washer shown in Figure 2-4.

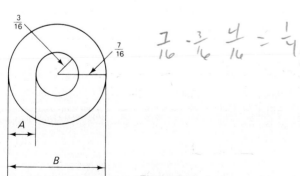

Figure 2-4

P.14

2-47. Find C given the cone pulleys shown in Figure 2-5.

2-48. Find D given the cone pulleys shown in Figure 2-5.

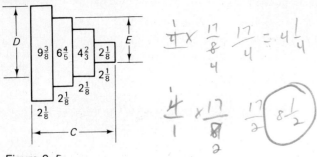

$\frac{1}{4} \times \frac{17}{8} \quad \frac{17}{4} = 4\frac{1}{4}$

$\frac{4}{1} \times \frac{17}{8} \quad \frac{17}{2} \quad \boxed{8\frac{1}{2}}$

Figure 2-5

2-49. Find E given the cone pulleys shown in Figure 2-5.

$\frac{6\frac{4}{5} + 2\frac{1}{8}}{2}$

$3\frac{2}{5} \qquad = 4\frac{37}{80}$

2-50. Find F given the concentric circles shown in Figure 2-6.

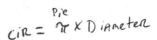

$CIR = \pi \times Diameter$ (Pie)

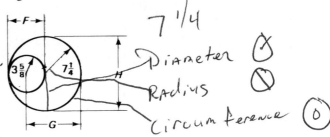

7 1/4

Diameter ○
Radius ⊘
Circumference ○

Figure 2-6

2-51. Find G given the concentric circles shown in Figure 2-6.

10 7/8

2-52. Find H given the concentric circles shown in Figure 2-6.

14 1/2

2-53. Find I given the eyebolt shown in Figure 2-7.

4 1/4

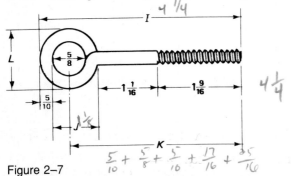

4 1/4

Figure 2-7

$\frac{5}{10} + \frac{5}{8} + \frac{5}{10} + \frac{17}{16} + \frac{25}{16}$

2-54. Find J given the eyebolt shown in Figure 2-7.

$\frac{5}{8} + \frac{5}{16} = 1\frac{1}{8}$

2-55. Find K given the eyebolt shown in Figure 2-7.

$\frac{1\frac{1}{16} + 1\frac{9}{16} + \frac{5}{10} + \frac{5}{8}}{2}$

$12\frac{5}{8} + \frac{5}{8}$

2-56. Find L given the eyebolt shown in Figure 2-7.

$\frac{5}{8} + 2(\frac{5}{10}) \frac{5}{5} = 1\frac{5}{8}$

2–57. Find *M* given the crankshaft shown in Figure 2–8.

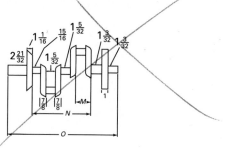

Figure 2–8

2–58. Find *N* given the crankshaft shown in Figure 2–8.

2–59. Find *O* given the crankshaft shown in Figure 2–8.

EXERCISE 2–5

Solve the following fraction and applied problems that involve operations with fractions. Remember to use the procedures you have learned.

2–60.
$$\frac{1}{4}$$
$$\frac{1}{2}$$
$$+\frac{5}{12}$$

2–61.
$$5\frac{1}{6}$$
$$+7\frac{3}{4}$$

2–62.
$$13\frac{2}{3}$$
$$7\frac{3}{8}$$
$$+ 9\frac{1}{4}$$

2–63. How many pounds of corn were produced if five plants produced the following amounts: $6\frac{1}{2}$, $4\frac{3}{4}$, $3\frac{7}{8}$, $4\frac{7}{32}$, and $5\frac{3}{16}$ pounds?

2–64. How much trim would you need for a window that is $6\frac{1}{2}$ by 8 feet?

2-65. $\frac{7}{12}$
       $-\frac{1}{3}$ $\frac{4}{12}$

       $\frac{3}{12} = \frac{1}{4}$

2-66. $17\frac{1}{16}$
       $-14\frac{3}{8}$

       $2\frac{11}{16}$

2-67. $16\frac{3}{8}$
       $-9\frac{15}{16}$

       $6\frac{7}{16}$

2-68. What is the difference in thickness of $\frac{5}{8}$-inch and $\frac{1}{2}$-inch sheetrock?

$\frac{5}{8} = \frac{5}{8}$
$-\frac{1}{2} = \frac{4}{8}$ = 1/8 Inches

2-69. How much of an 8-foot heater hose is left if you cut off a piece $3\frac{3}{4}$ feet long?

4 1/4 ft.

2-70. $32 \times 4\frac{1}{4} =$ 136

2-71. $\frac{1}{3} \times \frac{3}{8} \times \frac{64}{72} \times \frac{16}{54} =$ $\frac{8}{243}$

2-72. $2\frac{1}{3} \times 4\frac{3}{5} \times 2\frac{5}{30} =$

$23\frac{23}{90}$

2-73. Find the total length of eight studs if each is $7\frac{3}{4}$ feet long.

 62 ft.

2-74. If the travel time each way to work is $\frac{3}{4}$ hour, how many hours are spent in transit during a 5-day workweek?

$2(5) \times \frac{3}{4} = \frac{10 \times 3}{1 \times 4} = \frac{15}{2} = 7\frac{1}{2}$

2-75. $\frac{5}{6} \div \frac{3}{16} = \frac{16}{3} = \frac{40}{9} = 4\frac{4}{9}$

2-76. $7\frac{1}{5} \div 3\frac{3}{5} =$ 2

$\frac{36}{5} \times \frac{5}{18}$

2-77. $7\frac{1}{2} \div 1\frac{1}{4} =$

$\frac{15}{2} \times \frac{4}{5} = 6$

2–78. If $3\frac{1}{2}$ pizzas are to be divided among 12 persons, how much will each person receive?

$$\frac{3\frac{1}{2}}{\frac{12}{1}} = \frac{7}{2} \times \frac{1}{12} = \frac{7}{24}$$

2–79. A room $15\frac{1}{2}$ feet long is to have three electical outlets equally spaced. How far apart will they be?

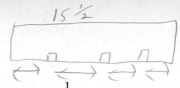

$$\frac{15\frac{1}{2}}{4} \quad 3\frac{7}{8}$$

2–80. $\dfrac{\frac{3}{5}}{2\frac{1}{8}} =$

2–81. $\dfrac{3\frac{1}{4} \div 4\frac{1}{8}}{4\frac{1}{4} \times 2\frac{7}{16}} = \dfrac{\frac{26}{33}}{\frac{1023}{64}}$

$$\frac{26}{33} \div \frac{663}{64} = \frac{26}{33} \cdot \frac{64}{663} = \frac{2}{33} \times \frac{64}{51}$$
$$= \frac{128}{1683}$$
$$.076054664\ldots$$

2–82. $\dfrac{1\frac{1}{3} \div 3}{2\frac{1}{16} + 1\frac{13}{32}} = \dfrac{4/9}{3\frac{15}{32}} = \dfrac{128}{999}$

2–83. What is the thickness of a pipe whose outside diameter is $3\frac{3}{8}$ inches and whose inside diameter is $2\frac{3}{4}$ inches?

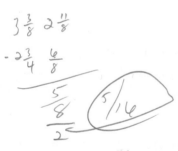

2–84. A "2 by 4" is actually $1\frac{5}{8}$ inches by $3\frac{5}{8}$ inches. How much is planed off *each* side of the $3\frac{5}{8}$-inch sides?

$$\frac{4 - 3\frac{5}{8}}{2} \quad \frac{3/8}{2} = 3/16$$

2–85. A piece of pipe is $14\frac{1}{2}$ inches long. How much is left after cutting off pieces measuring $2\frac{1}{2}$, $1\frac{1}{16}$, and $3\frac{5}{32}$ inches, allowing $\frac{1}{16}$ inch for each cut?

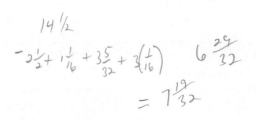

2–86. A taper has a diameter of $2\frac{5}{16}$ inches on one end and $1\frac{5}{8}$ inches at the other. Find the difference between the diameters.

$$11/16$$

2–87. The volume of a rectangular solid (block) is found by multiplying the length times the height. Find the volume of a block of concrete $11\frac{3}{4}$ inches by 8 inches by $6\frac{1}{2}$ inches.

2–88. A cubic foot contains approximately $7\frac{1}{2}$ gallons. How many gallons of gasoline would go into a $2\frac{3}{4}$-cubic-foot gas tank?

2–89. Find length A given the blueprint shown in Figure 2–9.

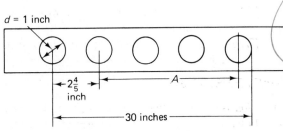

Figure 2–9

THINK TIME

There are many different ways to determine the lowest common denominator for fractions. Some of the ways have been shown. Perhaps you know other methods or can create new ones. Here is a great way to find the lowest common denominator, or LCD as it is known.

PROBLEM: Find the lowest common denominator for the fractions $\frac{1}{24} + \frac{3}{28} + \frac{7}{15}$.

SOLUTION: Place the denominators in an inverted division symbol as shown. Start by dividing the denominators by 2, a prime number. Place the results, 12 and 14, as well as 15, which could not be divided, in a new inverted division symbol. Repeat the process until no more divisions by 2 are possible, then divide by 3, another prime number, and continue this procedure until only prime numbers are results.

```
2|24   28  15
  2|12  14  15
    2|6   7  15
      3|3  7  15
         1  7   5
```

Then multiply the divisors and remaining prime numbers to obtain the lowest common denominator.

$$2 \times 2 \times 2 \times 3 \times 1 \times 7 \times 5 = 840$$

PROCEDURES TO REMEMBER

1. To reduce a fraction, divide the numerator and denominator by an equal quantity. *Example:*
$$\frac{12}{24} = \frac{12 \div 12}{24 \div 12} = \frac{1}{2}$$

2. To convert an improper fraction to a mixed number, divide the numerator by the denominator, write the quotient, and place the remainder over the denominator. *Example:*
$$\frac{5}{2} = 2\overline{)5} = 1\frac{1}{2}$$

3. To convert a whole number or a mixed number to an improper fraction, multiply the denominator by the whole number and add the original num-

ber to create the new numerator. *Examples:*
$2\frac{7}{8} = \frac{8 \times 2 + 7}{8} = \frac{23}{8}$ and $7 = \frac{7}{1}\left(\frac{5}{5}\right) = \frac{35}{5}$.

4. To multiply fractions:
   (a) Write whole numbers over 1.
   (b) Convert mixed numbers to improper fractions.
   (c) Reduce when possible. Do this by dividing the same quantity into the numerator and denominator (top into bottom or bottom into top).
   (d) Multiply the numerators and denominators.
   (e) Reduce to common fraction or to a mixed number.
5. When dividing fractions:
   (a) Write whole numbers over 1.
   (b) Convert mixed numbers to improper fractions.
   (c) Invert the fraction to the right of the division symbol.
   (d) Reduce when possible.
   (e) Multiply the numerators and the denominators.
   (f) Convert the result to a common fraction or a mixed number.
6. To find the lowest common denominator:
   (a) Determine the prime factors of each denominator.
   (b) Using the prime factors of the first denominator and the prime factors from other denominators not found in each of the other denominators, multiply these together. The result will be the lowest common denominator (LCD).
   (c) Check to make certain each denominator divides evenly into the common denominator.
7. When adding fractions:
   (a) Write the fractions to be added.
   (b) If they do not have a common denominator, find the lowest common denominator.
   (c) Change all fractions to equivalent fractions with the lowest common denominator.
   (d) Add all numerators.
   (e) Place the sum of the numerators over the common denominator.
   (f) Reduce to a common fraction or mixed number.
8. When subtracting fractions:
   (a) Write the fractions to be subtracted.
   (b) If they do not have a common denominator, find the lowest common denominator.
   (c) Change all fractions to equivalent fractions with the lowest common denominator.
   (d) Subtract the bottom numerator from the top numerator. If the bottom numerator is greater than the top numerator, borrow 1 from the whole number, convert the 1 to a fraction whose denominator is the same as the common denominator, add this to the numerator, then subtract the bottom numerator from the top numerator.
   (e) Place the difference over the lowest common denominator.
   (f) Reduce to a common fraction or mixed number.

CHAPTER SUMMARY

A *fraction* is a number that is used to compare a part to a whole. A fraction indicates one number divided by another number or a whole number over 1. Fractions are written to show the relationship of the part to the whole. An example of a fraction is $\frac{5}{8}$. The number below the line, 8, is called the *denominator* and tells the number of parts into which the whole has been divided. The number above the line, 5, is called the *numerator* and tells the number of parts of the whole.

The numerator and denominator are called the *terms* of the fraction. A fraction whose numerator is smaller than its denominator is called a *proper* fraction. A fraction whose numerator is larger than its denominator is called an *improper* fraction.

A whole number and a fraction written together is called a *mixed number*. A fraction that contains a fraction or fractions in the numerator or denominator is called a *complex* fraction. A *common denominator* is a number that is a multiple of the denominators of a set of fractions. The *lowest common denominator* is the smallest number that is a multiple of the denominators of a set of fractions. To *reduce* a fraction to its lowest terms, divide the numerator and the denominator by the same quantity. *Prime factors* are numbers that can be obtained by multiplying the quantities by 1.

CHAPTER TEST

T-2-1.
$$\frac{3}{4} \quad \frac{12}{16}$$
$$\frac{1}{2} \quad \frac{8}{16}$$
$$+\frac{5}{16}$$
$$\frac{25}{16} = 1\frac{9}{16}$$

T-2-2.
$$2\frac{2}{7}$$
$$+3\frac{1}{8}$$
$$5\frac{23}{56}$$

T-2-3.
$$15\frac{1}{2} \quad \frac{12}{24}$$
$$6\frac{1}{3} \quad \frac{8}{24}$$
$$+4\frac{3}{8} \quad \frac{9}{24}$$
$$25\frac{29}{24} = 26\frac{5}{24}$$

T-2-4.
$$\frac{15}{16}$$
$$-\frac{5}{8} \quad \frac{10}{16}$$
$$\frac{5}{16}$$

T-2-5.
$$2\frac{3}{4} \quad \frac{6}{8}$$
$$-1\frac{5}{8}$$
$$1\frac{1}{8}$$

T-2-6.
$$12\frac{1}{32} \quad \frac{33}{32}$$
$$-9\frac{7}{8} \quad \frac{28}{32}$$
$$2\frac{5}{32}$$

T-2-7. $\frac{5}{9} \times \frac{3}{10} = \frac{3}{18} = \frac{1}{6}$

T-2-8. $7\frac{1}{3} \times 2\frac{1}{2} = \frac{22}{3} \times \frac{5}{2} = \frac{55}{3} = 18\frac{1}{3}$

T-2-9. $4\frac{5}{16} \times 5\frac{1}{3} = 23$

T-2-10. $\frac{1}{8} \div \frac{3}{4} = \frac{4}{3} = \frac{1}{6}$

T-2-11. $\frac{25}{36} \div \frac{5}{9} = \frac{9}{8} \cdot \frac{5}{4} = 1\frac{1}{4}$

T-2-12. $5\frac{3}{8} \div 3\frac{1}{4} = 1\frac{17}{26}$

T-2-13. $\dfrac{9}{3\frac{2}{3}} = \frac{9}{1} \times \frac{3}{11} = \frac{27}{11} = 2\frac{5}{11}$

T-2-14. $\dfrac{\frac{5}{8} \times \frac{16}{45}}{\frac{3}{4} \div \frac{4}{3}} = \dfrac{2/9}{9/16} = \frac{32}{81}$

T-2-15. $\dfrac{\frac{2}{3} \times 7\frac{3}{4}}{\frac{1}{2} \times 37\frac{1}{16}} = \dfrac{5\frac{1}{6}}{18\frac{17}{32}}$

$$\dfrac{\frac{31}{6}}{\frac{593}{32}} = \frac{31}{6} \times \frac{32}{593} = \frac{496}{1779}$$

T-2-16. $18\frac{2}{3} \times 8\frac{1}{4} \div 72 =$

$154 \div 72 = 2\frac{5}{36}$

T-2-17. $6\frac{1}{4} \times 8\frac{1}{2} \times 5\frac{1}{2} \div 36 =$

$\frac{292\,3/6}{36} = 8\frac{67}{576}$

T-2-18. Find the dimension A for a set of eight steps (Figure 2-10).

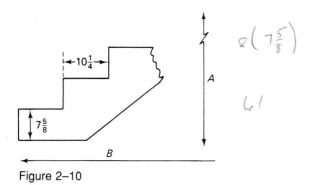

Figure 2-10

$8(7\frac{5}{8})$

61

T-2-19. Find the missing dimension B for a set of eight steps (Figure 2-10).

$10\frac{1}{4} \times 8 = 82$

T-2-20. A truck is loaded with bales of scrap iron, each weighing $\frac{1}{20}$ ton. If the total weight of the bales is $5\frac{1}{2}$ tons, how many bales are on the truck?

$\times \quad 5\frac{1}{2} \div \frac{1}{20} = 110\ TONS$

T-2-21. Eight inches of plastic tubing weighs $\frac{3}{4}$ ounce. How much will 2 inches of the tubing weigh?

$\frac{3}{4} \div 8 = \frac{3}{4} \times \frac{1}{8} = \frac{3}{32}$

$\frac{3}{32} \times \frac{2}{1} = \frac{3}{16}\ oz$

T-2-22. Twenty-four strips, each $1\frac{7}{8}$ inches wide, are to be ripped from a sheet of paneling. Twenty-four cuts are necessary and $\frac{1}{16}$ inch is lost on each cut. How wide must the paneling be?

$24(\frac{1}{16} + 1\frac{7}{8}) = 46\frac{1}{2}$

T-2-23. A master stonemason spends the following hours on different jobs: $4\frac{1}{2}, 8\frac{1}{2}, 1\frac{3}{4}, 2\frac{1}{4},$ and $5\frac{3}{4}$. At $\$20\frac{1}{2}$ per hour, what will the charge for labor be?

466.38

$\$466\frac{3}{8}$

Chapter 2 / Common Fractions

T-2-24. Partners Marge, Paul, and Steve agreed to share gains or losses in proportion to their respective investments. Marge invested $21,000; Paul, $28,000; and Steve, $35,000. If they are to share a gain of $126,000, how much should each receive?

T-2-25. A freshly molded plastic boat bumper weighed $14\frac{1}{2}$ kilograms. After being shaped, it weighed $12\frac{5}{8}$ kilograms. The shaping cost is $2 per kilogram, and the plastic removed in shaping can be sold for $$\frac{2}{3}$ per kilogram. Find the net cost of shaping the boat bumper.

DECIMALS

OBJECTIVES

1. To review the basic concepts and definitions of decimals.
2. To provide the opportunity to develop your skills working with decimals.
3. To use decimals to solve applied problems.

SELF-TEST

This test will evaluate your ability to work with decimals. If you are able to solve these problems, move on to percentages. Show your work; this will indicate areas where you may need to improve.

Read the following numbers and write them as figures.

S–3–1. Twelve and three tenths

12.3

S–3–2. Seventy-nine thousandths

.079

S–3–3. One hundred and one hundredth

100.01

S–3–4. Three and thirty-five thousandths

3.035

S–3–5. Two hundred sixty-five thousandths

0.265

S–3–6. Four hundred five and sixty thousandths

405.060

S–3–7. Forty-nine hundred-thousandths

.049

S–3–8. Four and twenty-four millionths

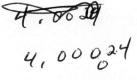

4.00024

Convert the following fractions to decimals.

S–3–9. $\frac{3}{5}$ .6

S–3–10. $\frac{3}{16}$ .1875

S–3–11. $\frac{7}{11}$ .63̄...

S–3–12. $\frac{1}{50}$ .02

S–3–13. $\frac{11}{28}$ .392857143...

S–3–14. $1\frac{8}{9}$ 1.8̄...

S–3–15. $\frac{230}{6}$ 38.3̄3̄...

Convert the following decimals to common fractions.

S–3–16. 0.75 $\frac{3}{4}$

S–3–17. 0.375 $\frac{375}{1000}$ $3/8$

S–3–18. 0.9375 (to the nearest 16th)

$\frac{15}{16}$

S–3–19. $0.34\frac{2}{3}$ $= \frac{\frac{104}{3}}{\frac{100}{1}}$

$\frac{\cancel{104}^{26}}{3} \times \frac{1}{\cancel{100}} = \frac{26}{75}$

S–3–20. What is the difference between $\frac{3}{4}$ and $\frac{5}{8}$ expressed as a decimal?

.125

S–3–21. There are seven steps in a flight of stairs. If each step is $8\frac{1}{4}$ inches high, find the distance between the floors. Express your answer as a decimal.

$7(8\frac{1}{4}) = 57\frac{3}{4}$ or 57.75 Inches

S–3–22. The wall of a building is made of cement blocks $9\frac{5}{8}$ inches thick, with plaster on one side $\frac{5}{16}$ inch thick. What is the decimal thickness of the wall?

$9\frac{15}{16} = 9.9375$ Inches

S–3–23. Northwest Airline's stock is selling for $25\frac{5}{8}$. What will be the cost of 40 shares?

$1025.

S–3–24. Number 12 wire is 0.081 inch in diameter. Find the diameter to the nearest 64th of an inch.

$\frac{.081}{1} \cdot \frac{64}{64} = \frac{5.184}{64} = \frac{5}{64}$

S–3–25. A piece of steel was $27\frac{7}{8}$ inches long. Four pieces of $4\frac{5}{8}$ inches were cut off. How long was the remaining portion if the saw blade was 0.012 inch wide?

$(4(4\frac{5}{8}) + 4(.012)) = 18.488$ 9.327
18½ − .012
$27\frac{7}{8} - 18.488 = 9.387$

INTRODUCTION

Decimals or decimal fractions are used whenever you make a purchase in a store. When you check the deductions on your paycheck, decimals are used. The dollars and cents used in everyday living require basic knowledge of decimals. As the metric system becomes more widely used in business and industry, you will need decimals to work effectively.

As more and more information is fed into computers to assist you in your occupation, you will need to know the mathematical language of the computer—decimals. Even though you may not work directly with computers, your occupation will rely heavily on the use of computers for time and money-saving results. Decimals are the language now and in the future.

DEFINITIONS

A *decimal* or *decimal fraction* is a fraction whose denominator is 10 or some multiple of 10, such as 10, 100, 1000, and so on. Examples of decimal fractions are $\frac{7}{10}$, $\frac{35}{100}$, $\frac{783}{1000}$, and $\frac{5,281}{10,000}$. When writing decimal fractions, the denominator is not placed below the numerator as it is with fractions. We indicate the denominator with a *decimal point* (.) in the numerator. The decimal point is placed so there are as many digits to the right of the decimal point as there are zeros in the denominator. Thus $\frac{47}{100}$ is written 0.47, $\frac{783}{1000}$ is written 0.783, and $\frac{532}{100}$ is written 5.32. The proof of this simple procedure may be shown by dividing. Thus

$$\begin{array}{r} .47 \\ 100\overline{)47.00} \\ \underline{40\ 0} \\ 7\ 00 \\ \underline{7\ 00} \\ 0 \end{array}$$

Chapter 3 / Decimals

Note: $\frac{47}{100}$ is called the fractional form and 0.47 is called the decimal form. A zero is placed to the left of the decimal point when there is no whole number. Thus 0.47 instead of .47, and 0.783 instead of .783. This is done to emphasize that the number is a decimal fraction.

When decimals are read, the decimal point may be called *point* or *and*. Thus 3.5 may be read "three point five" or "three and five tenths." The names of decimals are related to the denominator 10 or power of 10. Thus 0.3 is read "three tenths" and 0.74 is read "seventy-four hundredths." Names are assigned to each *place* value to the left and right of the decimal point. Study Figure 3–1. The two columns on the left name the number and its word form. The third column shows how the multiplication of ten or tenths produces the number that appears in the first column. The multiplication process shows the *power of 10*. The fourth column indicates the power of 10 when the number is written in a simpler form called the *exponential form*. Understanding the two columns on the right will help when you solve problems.

Thus the number 69,373.56354 is read "sixty-nine thousand, three hundred seventy-three and fifty-six thousand, three hundred fifty-four one hundred thousandths." If no decimal point is given, it is understood the decimal point is to the right of the last digit. Thus 12 = 12.0 and 528 = 528.0.

Number	Name	Powers-of 10 Form	Exponential Form
1,000,000	one million	$10 \times 10 \times 10 \times 10 \times 10 \times 10$	10^6
100,000	one hundred thousand	$10 \times 10 \times 10 \times 10 \times 10$	10^5
10,000	ten thousand	$10 \times 10 \times 10 \times 10$	10^4
1000	one thousand	$10 \times 10 \times 10$	10^3
100	one hundred	10×10	10^2
10	ten	10	10^1
1	one	1	10^0
0.1	one tenth	$\frac{1}{10}$	$\left(\frac{1}{10}\right)^1$ or 10^{-1}
0.01	one hundredth	$\frac{1}{10} \times \frac{1}{10}$	$\left(\frac{1}{10}\right)^2$ or 10^{-2}
0.001	one thousandth	$\frac{1}{10} \times \frac{1}{10} \times \frac{1}{10}$	$\left(\frac{1}{10}\right)^3$ or 10^{-3}
0.0001	one ten-thousandth	$\frac{1}{10} \times \frac{1}{10} \times \frac{1}{10} \times \frac{1}{10}$	$\left(\frac{1}{10}\right)^4$ or 10^{-4}
0.00001	one hundred-thousandth	$\frac{1}{10} \times \frac{1}{10} \times \frac{1}{10} \times \frac{1}{10} \times \frac{1}{10}$	$\left(\frac{1}{10}\right)^5$ or 10^{-5}
0.000001	one millionth	$\frac{1}{10} \times \frac{1}{10} \times \frac{1}{10} \times \frac{1}{10} \times \frac{1}{10} \times \frac{1}{10}$	$\left(\frac{1}{10}\right)^6$ or 10^{-6}

Figure 3–1 Place values for decimals

EXERCISE 3-1

Write the following in as fractions with a denominator of 10 or power of 10.

3-1. 0.09

3-2. 0.085

3-3. 521.14

3-4. 16.141

Write the following as decimals.

3-5. Forty hundredths

.40

3-6. Eight tenths

3-7. Seven and four tenths

3-8. Five and thirty-two thousandths

3-9. Two and two hundredths

3-10. Thirty-six and twenty-four thousandths

Write the following decimals in words.

3-11. 0.909

. 9 hundred nine thousandths

3-12. 3.005

Chapter 3 / Decimals

3–13. 5.0055

3–14. 168.09

3–15. 16,000.00016

Write the following in fractions with denominators of 10 or power of 10.

3–16. Thirty-three and three tenths

$33 \frac{3}{10}$

3–17. Five hundred and four hundred-thousandths

$500 \frac{4}{100,000}$

3–18. Three and five hundred thirty-five ten thousandths

$3 \frac{535}{10,000}$

3–19. Seventy-seven hundred-thousandths

$\frac{77}{100,000}$

3–20. Nine hundred thirty-four ~~thousand~~ and six ten-thousandths

$934 \frac{6}{10,000}$

CHANGING FRACTIONS TO DECIMALS

A common fraction may be converted to a decimal by dividing the denominator into the numerator.

Example 3–1.

  PROBLEM: Convert $\frac{3}{4}$ to a decimal.

  SOLUTION: Divide the denominator (4) into the numerator (3).

```
      .75
  4)3.00
    2 8
    ---
      20
      20
```

Therefore, the fraction $\frac{3}{4}$ equals the decimal 0.75.

$$\frac{3}{4} = 0.75$$

40   Section One / Arithmetic

Example 3-2

PROBLEM: Convert $\frac{9}{16}$ to a decimal.

SOLUTION: Divide the denominator (16) into the numerator (9).

$$\begin{array}{r} .5625 \\ 16\overline{)9.0000} \\ \underline{8\,0} \\ 1\,00 \\ \underline{96} \\ 40 \\ \underline{32} \\ 80 \\ \underline{80} \end{array}$$

Thus $\frac{9}{16} = 0.5625$.

Example 3-3

PROBLEM: Convert the mixed number $15\frac{5}{6}$.

SOLUTION: Since 15 is a whole number, it may be placed to the left of the decimal point; then divide the numerator (5) by the denominator (6).

$$\begin{array}{r} .8333 \\ 6\overline{)5.0000} \\ \underline{4\,8} \\ 20 \\ \underline{18} \\ 20 \\ \underline{18} \\ 20 \\ \underline{18} \\ 2 \end{array}$$

Note: Regardless of how long you continue to divide, you will always obtain a 3. This is called a *repeating decimal*. It may be written three ways: by placing a dot above the 3: $\dot{3}$; by placing a short line above the 3: $\bar{3}$; or by placing three dots immediately after the last 3: 3 . . . . Thus

$$15\frac{5}{6} = 15.8\dot{3} = 15.8\bar{3} = 15.83\ldots$$

EXERCISE 3-2

Convert the following fractions to decimals.

3-21. $\frac{3}{10}$

3-22. $\frac{5}{8}$

3-23. $\frac{5}{16}$

3-24. $\frac{49}{100}$

3-25. $\frac{1}{9}$

3-26. $\frac{2}{32}$

3-27. $\frac{21}{32}$

3-28. $\frac{37}{8}$

3-29. $\frac{31}{64}$

3–30. Write 5 cents as a decimal part of a dollar.

CHANGING DECIMALS TO FRACTIONS

In business and industry, it is often necessary to change decimals to fractions. One common need is to change decimals to fractions with denominators of fourths, eighths, sixteenths, thirty-seconds, or sixty-fourths. These changes need to be made rapidly. Thus there are *tables of decimal equivalents*. The exact value is frequently not on the table. Therefore, the decimal must be rounded off to the nearest fraction, depending on the purpose of the conversion. These conversions will become more and more important as industry and business become more international and the metric system the measurement system.

To convert any decimal to a fraction, simply write the decimal as a fraction. For example, 0.5 may be written as $\frac{5}{10}$, 0.79 as $\frac{79}{100}$, and 0.329 as $\frac{329}{1000}$. These fractions may be reduced by dividing the numerator and the denominator by the same quantity. In decimal fractions, the denominator is divisible by the numbers 2, 5, and 10.

Example 3–4

PROBLEM: Convert 0.65 to a fraction and reduce to lowest terms.

SOLUTION: Convert the decimal to its decimal fraction form. Write as a fraction.

$$0.65 = \frac{65}{100}$$

Reduce the fraction by dividing the numerator and denominator by 5.

$$\frac{\cancel{65}^{13}}{\cancel{100}_{20}} = \frac{13}{20}$$

Thus $0.65 = \frac{65}{100}$.

Example 3–5

PROBLEM: Convert 3.875 inches to a mixed number and reduce to lowest terms.

SOLUTION: Convert the decimal to its decimal fraction form.

$$3.385 = 3\frac{875}{1000}$$

Reduce the fraction by dividing the numerator and denominator by 25.

$$3\frac{\cancel{875}^{13}}{\cancel{1000}_{40}} = 3\frac{35}{40}$$

Reduce the fraction again by dividing the numerator and denominator by 5.

$$3\frac{\cancel{35}^{7}}{\cancel{40}_{8}} = 3\frac{7}{8}$$

Thus, $3.875 = 3\frac{7}{8}$. Remember, several steps may be necessary to reduce a fraction to its lowest form.

Example 3–6

PROBLEM: Convert 0.316 to the nearest sixteenth.

SOLUTION: Place 0.316 over 1 and multiply the numerator and denominator by 16.

$$\frac{0.316}{1} \times \frac{16}{16} = \frac{0.316 \times 16}{1 \times 16} = \frac{5.056}{16}$$

Round to the nearest sixteenth.

$$\frac{5.056}{16} = \frac{5}{16}$$

Thus $0.316 = \frac{5}{16}$, to the nearest sixteenth.

Example 3–7

PROBLEM: Convert $0.83\frac{1}{3}$ to a common fraction.

SOLUTION: Convert the decimal to a decimal fraction.

$$0.83\frac{1}{3} = \frac{83\frac{1}{3}}{100}$$

Convert the numerator to an improper fraction by multiplying 83 by 3 and adding 1.

$$0.83\frac{1}{3} = \frac{83\frac{1}{3}}{100} = \frac{\frac{250}{3}}{\frac{100}{1}}$$

Then do the indicated division.

$$\frac{\frac{250}{3}}{\frac{100}{1}} = \frac{250}{3} \div \frac{100}{1} = \frac{\overset{5}{\cancel{250}}}{3} \times \frac{1}{\underset{2}{\cancel{100}}} = \frac{5}{6}$$

Thus $0.83\frac{1}{3} = \frac{5}{6}$.

EXERCISE 3–3

Convert the following decimals and reduce to lowest terms.

3–31. 0.7   7/10

3–32. 0.16   4/25

3–33. 0.625   5/8

3–34. 0.375   3/8

3–35. $0.72\frac{8}{11}$ 18/25

3–36. 0.0125   1/80

3–37. Convert 0.34 to the nearest eighth.

$.34 \times \frac{8}{8}$ $\frac{2.72}{8} = \frac{3}{8}$

3–38. Convert 0.367 to the nearest sixteenth.

$\frac{.367}{1} = \frac{x}{16}$ $x = 16 \cdot .367$

$x = 5.872$

$\frac{5.872}{16} = \frac{6}{16}$

3–39. Convert 8.096 to the nearest thirty-second.

$8.\left[.096 \times \frac{32}{32}\right]$ $\frac{3.072}{32}$

$= 8\,\frac{3}{32}$

3–40. Convert 12.5326 to the nearest sixty-fourth.

$\frac{12.5326}{1} \times \frac{64}{64} = \frac{802.0864}{64} = \frac{802}{64} =$

$12\,\frac{17}{32} = 12\,\frac{34}{64}$

Chapter 3 / Decimals

3-41. Convert $\frac{7}{8}$ inch to a decimal.

3-42. Convert $40\frac{1}{2}$ inches to decimal form in feet to the nearest thousandth. (12 inches = 1 foot.)

3-43. Find the inside diameter, in decimal form, of a pipe with an outside diameter of $6\frac{1}{2}$ inches if the pipe is $\frac{3}{16}$ inch thick.

3-44. A board 6 inches wide is to be finished to seven eighths of its original width. Express the new width as a decimal.

3-45. What is the dollar value of stock with a market value of $75\frac{3}{8}$?

3-46. A machine produced 495 outboard motor shear pins in 8.25 minutes. Find the average number of shear pins produced per minute.

3-47. A piece of marine plywood is made up of one piece of $\frac{1}{4}$-inch wood and two pieces of $\frac{1}{16}$-inch wood. Find the total thickness of the plywood in decimal form.

3-48. A steel shaft is $1\frac{5}{8}$ inches in diameter. Using a lathe, how deep a cut must be made to reduce the diameter to 1.594 inches?

3-49. A mechanic wants to set the spark plug gap at 0.035. Express the gap as a fraction to the nearest sixty-fourth.

3-50. A blueprint calls for a 0.8365-inch hole or slightly larger. The available drills are sized in thirty-seconds of an inch. What drill should be used?

THINK TIME

It is important to know decimal and fractional equivalents. Figure 3–2 shows a mixture of decimals and fractions. Connect the pairs of equivalents with straight lines. When you have completed this, you will have a five-pointed star.

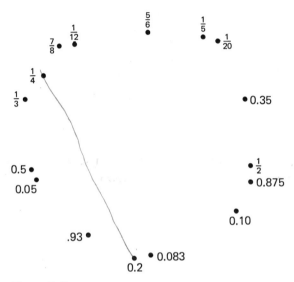

Figure 3–2

PROCEDURES TO REMEMBER

1. To change a fraction to a decimal:
   (a) Set up the division of the denominator into the numerator.
   (b) Place the decimal point.
   (c) Divide, carrying the division as many places as required.
2. To change a decimal to a fraction:
   (a) Write the decimal in its decimal fraction form in a power of 10.
   (b) Reduce the fraction to its lowest terms.
3. To change a decimal to a fraction with a specific denominator (eighth, sixteenth, etc.):
   (a) Write the decimal over 1.
   (b) Multiply the numerator and denominator by the specific denominator (eighth, sixteenth, etc.)
   (c) Round the numerator to the nearest whole number.

CHAPTER SUMMARY

A *decimal fraction* is a fraction whose denominator is 10 or a power of 10. The *decimal point* (.) separates the whole number from the decimal part of a number. The word *point* or *and* is used to indicate a decimal point when reading decimal quantities. A value that continues to repeat when one quantity is divided into another is called a *repeating decimal*. Example: $\frac{1}{3} = 0.\overline{3}$ or $\dot{3}$ or 0.333 . . . . A repeating decimal is indicated by a bar or dot over the repeating digit or by three dots after the repeating digit. Decimal fractions may be written in two ways: *fractional form* $\left(\frac{37}{100}\right)$ or in *decimal form* (0.37).

CHAPTER TEST

Read the following numbers and then write them as figures.

T–3–1. Seven and five hundredths

T–3–2. Eighty-four thousandths

T–3–3. One thousand and one thousandth

T–3–4. Seven and sixty-six thousandths

Chapter 3 / Decimals

T-3-5. Three hundred and thirty-three thousandths

T-3-6. Three thousand and six thousandths

T-3-7. Fifty-five and ten-thousandths

T-3-8. Three and three thousand one hundred twenty-five hundred thousandths

Convert the following fractions to decimals.

T-3-9. $\dfrac{4}{5}$

T-3-10. $\dfrac{11}{32}$

T-3-11. $\dfrac{7}{9}$

T-3-12. $\dfrac{2}{25}$

T-3-13. $\dfrac{26}{27}$

T-3-14. $3\dfrac{3}{11}$

T-3-15. $\dfrac{25}{16}$

Convert the following decimals to fractions.

T-3-16. 0.15

T-3-17. 0.1875

T–3–18. Convert 0.844 to the nearest thirty-second.

T–3–19. Convert $0.44\frac{4}{9}$ to a common fraction.

T–3–20. Find the decimal difference between $\frac{7}{8}$ and $\frac{5}{6}$ (to the nearest thousandth).

T–3–21. A machinist turned a $\frac{27}{64}$-inch-diameter shaft from a piece of stock $\frac{1}{2}$ inch in diameter. Express the difference in diameter as a decimal.

T–3–22. A plumber made a run of pipe consisting of three pieces whose lengths were $15\frac{7}{16}$ inches, $10\frac{3}{8}$ inches, and $18\frac{7}{8}$ inches. Express the total length of pipe as a decimal.

T–3–23. An interactive laser disk stock is selling for $\$48\frac{7}{8}$. What will 55 shares of the stock cost?

T–3–24. A blueprint calls for a 0.84375-inch opening. To the nearest thirty-second, what drill should be used?

T–3–25. When plastic cures, it reduces in size by $\frac{1}{250}$. Find the length, in decimal form, that a 30-foot plastic pipe would be after curing.

Chapter 3 / Decimals

4

PERCENTAGES

OBJECTIVES

1. To review the skills to:
   (a) Convert common fractions to decimals.
   (b) Convert decimals to common fractions.
   (c) Convert decimals to percents.
2. To develop the skills to:
   (a) Find the percent of a number.
   (b) Compare one number to another using percent.
   (c) Solve applied problems.
3. To develop skills solving percent problems using the hand-held calculator.

SELF-TEST

The skills to solve percent problems are similar to the skills used doing decimals. Your results from this test may indicate you need to review this chapter. You can identify areas to review by checking your work.

S–4–1. Convert $\frac{11}{16}$ to a decimal.

.06875

S–4–2. Convert $5\frac{7}{8}$ to a decimal.

5.875

S–4–3. Convert 35% to a fraction.

7/20

S–4–4. Convert $37\frac{1}{2}\%$ to a fraction.

3/8

S–4–5. Convert 0.37 to a percent.

37%

S–4–6. Convert 0.08 to a percent.

8%

S–4–7. Convert 0.005 to a percent.

.5%

S–4–8. Convert $\frac{1}{2}\%$ to a fraction.

.5%/200

S-4-9. Convert $34\frac{2}{3}\%$ to a fraction.

$\frac{26}{75}$

S-4-10. Convert 0.115 to a percent.

11.5

S-4-11. $63\% = \frac{?}{100}$ 63

S-4-12. Convert $\frac{7}{10}$ to a percent.

.007   70%

S-4-13. $\frac{6}{25} = \frac{?}{100}$ 24

S-4-14. What is 30% of 50?

15

S-4-15. Find 39% of 1992.

776.88

S-4-16. If $3\frac{1}{2}\%$ of the transistors manufactured are defective, how many would be defective if 5000 were produced?

175 transistors defective

S-4-17. If fertilizer is 12% nitrogen, how much nitrogen is in a 22-pound bag?

2.64 LB.

S-4-18. What is $6\frac{3}{4}\%$ of $2800?

$189

S-4-19. 800 is 12% of what number?

6666.6

S-4-20. What is the sales tax on a new automobile costing $13,988.63 at a tax rate of 6%?

$839.32

S-4-21. If 6.3% of your salary is deducted for the retirement fund, how much would be deducted from a gross weekly salary of $488?

$30.74

S-4-22. What is the total interest on an automobile loan of $3500 if the interest rate is $9\frac{1}{4}$% per year for 3 years?

$971.25

S-4-23. What is the total interest on a $400 life insurance loan at $5\frac{1}{4}$% each year for 12 years?

$252.00

S-4-24. A transistor radio was marked down from $32 to $24. Find the percent of discount.

25%

S-4-25. If the attendance at a church was 2300 at the beginning of the year, what would it be at the end of the year if it increased $12\frac{1}{2}$%?

2587.5

INTRODUCTION

Percentages are used to indicate a relationship between numbers. Frequently, percentages are used to show the rate of change, the interest, and determine tax. Changes are indicated in percentages for increases or decreases in wages, crime, cost of living, production, natural resources, and so on. Thinking customers study the rates for interest on loans, carrying charges, and mortgages. Many dollars are saved by selecting lending agencies that offer the lowest rate of interest. Skills working with percentages will be helpful both on the job and for personal use.

DEFINITIONS

Percent represents the number of parts out of 100 or the comparison of a number to another number. Generally, we compare a quantity or number by placing it as a numerator over the denominator, 100. The word *percent* is derived from the Latin word *percentum* and means "by the hundred." The symbol used for percent is %. Basically, there are three types of percentage problems. Examples are as follows: (a) What is 20% of 540? (b) Thirty-seven is what percent of 97? (c) Sixty-three is 25% of what number? All problems concerning percentages may be traced back to these examples.

Percent problems involve three things: the *base,* the *rate of percent,* and the *percentage.* These concepts can best be explained by an example. A man earns $200 a week and saves 30% of what he earns. Thus he saves $60 a week. In this problem, $200 is the base, 30% is the rate of percent, and $60 is the percentage. To find the percentage or amount saved, take 30% of $200, which is $60; 30% of $200 is $60 because 30% of $200 means 0.30 × $200 = $60.

The following formulas are to be used to find percentage, base, and rate.

1. To find the *percentage,* use the formula

   percentage = base times rate $p = b \times r$

2. To find the *base,* use the formula

   base = percentage divided by rate $b = \frac{p}{r}$

3. To find the *rate,* use the formula

$$\text{rate} = \text{percentage divided by base} \qquad r = \frac{p}{b}$$

Sometimes it is difficult to determine if a number represents the rate, base, or percentage. Some helpful hints concerning percentages are as follows:
1. The rate always has a percent sign or the word "percent" with it.
2. The base is the number that follows the word "of" in a word problem and refers to the entire quantity.
3. The percentage is the result of finding the rate when the base is given.

CONVERTING FRACTIONS, DECIMALS, AND PERCENTS

The rate of percent is converted to a decimal for easier computation. To convert a percent to a decimal, remove the percent sign and move the decimal point two places to the left. This is equivalent to dividing the percent by 100.

To convert a decimal to a percent, move the decimal point two places to the right and write the percent sign, %, after the number. This is equivalent to multiplying the decimal by 100.

To change a fraction to a percent, convert the fraction into a decimal, then move the decimal two places to the right and add the % symbol.

To change a percent to a fraction, remove the percent sign, place the number over 100, then reduce to lowest terms. To understand these procedures better, work through the following examples.

Example 4–1

PROBLEM: Change 15% to a decimal.

SOLUTION: Remove the percent sign and move the decimal point two places to the left.

$$15\% = 0.15$$

Example 4–2

PROBLEM: Change 0.075 to a percent.

SOLUTION: Move the decimal point two places to the right and add the percent symbol.

$$0.075 = 7.5\% \quad \text{or} \quad 7\frac{1}{2}\%$$

Example 4–3

PROBLEM: Change 7 over 8 to a percent.

SOLUTION: Divide 8 into 7.

$$\frac{7}{8}\% = 0.875$$

Move the decimal point two places to the right and add the percent sign.

$$0.875 = 87.5\% \quad \text{or} \quad 87\frac{1}{2}\%$$

Example 4–4

PROBLEM: Change $62\frac{1}{2}\%$ to a fraction.

SOLUTION: Remove the percent sign and place $62\frac{1}{2}$ over 100.

$$62\frac{1}{2}\% = \frac{62\frac{1}{2}}{100} = \frac{62.5}{100} = \frac{625}{1000}$$

Reduce to lowest terms.

$$\frac{\overset{25}{\overset{\cancel{625}}{\cancel{1000}}}}{\underset{40}{}} = \frac{\overset{5}{\cancel{25}}}{\underset{8}{\cancel{40}}} = \frac{5}{8}$$

Example 4–5

PROBLEM: Change $16\frac{2}{3}\%$ to a fraction.

SOLUTION: Remove the percent sign and place $16\frac{2}{3}$ over 100.

$$16\frac{2}{3}\% = \frac{16\frac{2}{3}}{100}$$

Change the numerator and denominator to improper fractions.

$$\frac{16\frac{2}{3}}{100} = \frac{\frac{50}{3}}{\frac{100}{1}}$$

Rewrite the complex fraction as a division-of-fractions problem.

$$\frac{\frac{50}{3}}{\frac{100}{1}} = \frac{50}{3} \div \frac{100}{1}$$

Invert and multiply.

$$\frac{50}{3} \div \frac{100}{1} = \frac{50}{3} \div \frac{1}{100} = \frac{1}{6}$$

Therefore, $16\frac{2}{3}\% = \frac{1}{6}$.

EXERCISE 4–1

Convert the following percents, decimals, and fractions as indicated.

4–1. Convert 37% to a decimal.

.37

4–2. Convert 0.865 to a percent.

4–3. Convert 16.5 to a percent.

16.5%
1650%

4–4. Convert 0.173 to a percent.

4–5. $\frac{1}{8}$ equals what percent?

.001
.125%

4–6. Convert 0.3% to a fraction.

4–7. Convert 25.5% to a decimal.

.255

4–8. Convert $33.\overline{3}\%$ to a fraction.

4–9. Convert 37.5% to a fraction.

3/8

4–10. Convert 0.08 to a percent.

4–11. 0.0091 equals what percent?   wrong

9.1%   .0091%
       91%

4–12. Convert $\frac{4}{7}$ to a percent.

4–13. Convert $\frac{3}{8}$ to a percent.

.375
37.5%

4–14. Convert $28\frac{4}{7}$% to a fraction.

28 4/7 ÷ 100 =

2/7

4–15. Convert 0.036 to a percent.

3.6%

4–16. Convert $\frac{3}{10}$ to a decimal.

.3

4–17. $12\frac{1}{2}$% equals what decimal value?

.125

4–18. Convert $66\frac{2}{3}$% to a fraction.

4–19. Convert $\frac{5}{6}$ to a percent.

.83.3̄%

4–20. Convert $\frac{5}{12}$ to a percent.

APPLICATIONS

It is important to be able to apply your skills with percent to everyday problems. The next group of examples will give you an opportunity to develop your skills working with percents.

Example 4–6

PROBLEM: Find 30% of 360 days.

SOLUTION: Convert 30% to a decimal.

30% = 0.30

Since "of" means multiply, multiply as shown.

360 days × 0.30 = 108 days

Thus 30% of 360 days is 108 days.

Example 4–7

PROBLEM: Find 121% of $1500.

SOLUTION: Convert 121% to a decimal.

121% = 1.21

Multiply as shown.

$1500 × 1.21 = $1815

Thus 121% of $1500 is $1815.

Chapter 4 / Percentages

Example 4-8

PROBLEM: Determine the selling price of a small motorcycle if the cost to the dealer is $1224 and the profit is 30% of the cost.

SOLUTION: Change 30% to a decimal.

$$30\% = 0.30$$

Multiply $1224 times 0.30 to find the profit.

$$\$1224 \times 0.30 = \$367.20$$

Then add the profit to the dealer's cost to find the selling price.

$$\begin{array}{ll} \$1224.00 & \text{cost} \\ +\ \ \ 367.20 & \text{profit} \\ \hline \$1591.20 & \end{array}$$

ALTERNATIVE SOLUTION: As the selling price is 100% of the dealer's cost plus 30% profit, we could add these percents and then multiply the dealer's cost by this percent. Follow these procedures to solve the problem. Add 100% and 30%.

$$\begin{array}{r} 100\% \\ +\ \ 30\% \\ \hline 130\% \end{array}$$

Convert 130% to a decimal.

$$130\% = 1.30$$

Multiply the dealer's cost ($1224) by 1.30 to find the selling price:

$$\text{selling price} = \$1224 \times 1.30 = \$1591.20$$

Example 4-9

PROBLEM: Find the cost of a pair of snow tires that were priced at $96 but were reduced 20%.

SOLUTION: Discounts are given as a percent off the original price. To find the discount, multiply the original price by the rate of discount (20%), then subtract the discount ($19.20) from the original selling price ($96) to find the discounted price ($78.80). Change 20% to a decimal.

$$20\% = 0.20$$

Multiply the original selling price ($96) by the decimal (0.20):

$$\text{discount} = \$96 \times 0.20 = \$19.20$$

Subtract the discount ($19.20) from the original price of $56.

$$\begin{array}{ll} \$96.00 & \text{original price} \\ +\ 19.20 & \text{reduction} \\ \hline \$76.80 & \end{array}$$

ALTERNATIVE SOLUTION: The reduced price is actually 80% of the original price. Subtract 20% from 100% to get this.

$$100\% - 20\% = 80\%$$

Change 80% to a decimal.

$$80\% = 0.80$$

Multiply 0.80 times the original price ($96) to determine the reduced price:

$$\text{reduced price} = \$96 \times 0.80 = \$76.80$$

EXERCISE 4-2

4-21. What is 42% of 536?

225.12

4-22. What is 9% of 15?

1.35%

4-23. What is 275% of 828?

2277.

4-24. Find 343% of 606?

2078.58%

4-25. Find 9% of 180?

16.2

4-26. What is 123% of 41?

50.43

4-27. Find 76% of $584.

443.84

4-28. What is 0.4% of 5280?

21.12

4-29. What is 325% of 60?

195

4-30. What is 5.5% of $18?

$0.99

4-31. The profit on an electric calculator is 55%. What is the selling price if the cost is $18?

27.90

4-32. Find the state sales tax on a new car priced at $14,995 if the tax rate is 4%.

$599.8

4-33. What number is 6% of $580?

$34.8

4-34. A dress costing $27 is reduced 15%. What is the reduced price?

$22.95

4–35. A football team has an 87.5% winning percentage. How many games of a 16-game schedule did the team win?

14 games

4–36. A grain buyer bought wheat at $5.40 a bushel and sold it at an 8% loss. What was the selling price of the wheat?

$4.97

4–37. The 6% sales tax on a pontoon boat amounted to $473.70. What was the cost of the boat?

$? = \frac{473.70}{.06}$

4–38. The city sales tax on a color television set was $28 for an $800 set. Determine the sales tax rate.

$28/800 = 3.5\%$

4–39. A mixture for concrete consists of 3 parts sand, 2 parts gravel, and 1 part cement. What is the percent of cement in the mixture?

$16.\overline{6}\%$

4–40. A small business building is insured for 90% of its value at an insurance rate of $\frac{1}{4}$% per year. The building is valued at $140,000. Determine the annual premium.

315.00

INTEREST

Interest is money paid for the use of money. It is determined by a percent of the amount of money being loaned. Generally, interest is determined on the basis of 1 year. Banks and other lending agencies pay interest on the amount placed in the bank, the rate of interest being a percent per year. The base is called the *principal*, or the amount of money loaned. The *percentage* is the interest paid. Interest is found by multiplying the rate *times* the principal *times* the amount of time, which is generally by the year but could be a fractional part of a year. For example, 5 years 9 months would be written $5\frac{9}{12}$ or 5.75 years.

The formula to find interest is $i = prt$. Sales taxes and commissions are calculated in the same manner.

Example 4–10

PROBLEM: Find the interest on $2500 for 1 year at $12\frac{1}{2}$% per year.

SOLUTION: Interest equals principal times the rate times the time, or $i = prt$. Identify the given quantities.

$$i = \$2500 \times 12\frac{1}{2}\% \times 1 \text{ (year)}$$

Change $12\frac{1}{2}$% to a decimal.

$$12\frac{1}{2}\% = 0.125$$

Multiply the principal times the rate times the time.

$$i = prt$$
$$= \$2500 \times 0.125 \times 1$$
$$= \$312.50$$

The interest on $2500 at $12\frac{1}{2}$% for 1 year is $312.50.

Example 4–11

PROBLEM: Interest of $120 is paid at a rate of 5% per year. How much money (principal) is in the account?

SOLUTION:

$$\text{Principal} = \frac{\text{interest (\$120)}}{\text{rate (5\%)} \times \text{time (1 year)}} \quad \text{or}$$

$$p = \frac{i}{rt}$$

Identify the given quantities.

$$p = \frac{i}{rt}$$
$$= \$\frac{120}{5\% \times 1}$$

Change 5% to a decimal.

$$5\% = 0.05$$

Divide the percentage ($120) by the rate times the time to find the principal.

$$p = \$\frac{120}{(0.05)(1)}$$
$$= \$2400$$

Thus $2400 is the principal in the savings account.

Example 4–12

PROBLEM: Find the rate of interest when $7955 is paid for 1 year on a mortgage of $86,000.

SOLUTION:

$$\begin{array}{c}\text{Rate} \\ \text{of} \\ \text{interest}\end{array} = \frac{\text{interest (\$7955)}}{\text{principal (\$86,000)} \times \text{time (1 year)}} \quad \text{or}$$

$$r = \frac{i}{pt}$$

Identify the given quantities.

$$i = \$7955 \quad \text{and} \quad p = \$86,000$$
$$r = \frac{\$7955}{(\$86,000)(1)}$$

Divide the interest ($7955) by the principal ($86,000) times the time (1 year).

$$r = \frac{\$7955}{(\$86,000)(1)}$$
$$= 0.0925$$

Change the decimal (0.0925) to a percent.

$$0.0925 = 9.25\%$$

The rate of interest on the mortgage is 9.25%.

Example 4–13

PROBLEM: On a $10,000 municipal bond, the total interest received is $5300. If the rate of interest is $6\frac{5}{8}\%$, how many years has the interest been paid to total $5300?

SOLUTION:

$$\text{Time} = \frac{\text{interest (\$5300)}}{\text{principal (\$10,000)} \times \text{rate}\left(6\frac{5}{8}\%\right)} \quad \text{or}$$

$$t = \frac{i}{pr}$$

Identify the given quantities.

$$t = \frac{\$5300}{(\$10,000)\left(6\frac{5}{8}\%\right)}$$

Change $6\frac{5}{8}\%$ to a decimal.

$$6\frac{5}{8} = 0.06625$$

Evaluate

$$t = \frac{\$5300}{(\$10,000)(0.06625)}$$
$$= 8 \text{ years}$$

Thus interest of $6\frac{5}{8}\%$ has been paid for 8 years on the $10,000 bond.

EXERCISE 4-3

Solve the following applied percent problems to improve your skills with practical problems.

4-41. Find the interest on $580 at 8% for 1 year.

$46.4

4-42. Find the interest on $1250 at $9\frac{1}{2}$% for 1 year.

4-43. Find the interest on $17,500 at $7\frac{1}{4}$% for 20 years.

$25,375

4-44. Find the interest on $4050 at 12.5% for $3\frac{1}{2}$ years.

506.25 = $1771.88

4-45. Find the monthly payments on a $13,500 automobile at $10\frac{1}{2}$% for 3 years.

1417.5
+ 13,500
12)14,917.5

+375.00
118,...
493,125

4-46. Find the rate of discount for an AM–FM radio priced at $49.95 that was reduced to $39.95 for a special sale.

20%

4-47. Find the interest on a $250 savings account for 5 years 9 months at $7\frac{1}{2}$%.

18.75
$107.81

4-48. If an electric bill is paid within 15 days, a discount of $1\frac{1}{2}$% is given. What would a customer pay on a $33 electric bill if it was paid within 15 days?

50.
32.5¢

4-49. A dealer allowed $2875 for a used car. He then sold the car for $3000. What was his percent of gain?

The Difference would be $125.00 = 2875
4.35%

4-50. Find the rate if the principal is $8000 and the interest for 1 year is $400.

5% 8000)400

4–51. What is the rate of interest if the principal is $3300 and the interest received after 8 years is $924?

3.5%

4–52. Kay Cobb's Furniture offered $\frac{1}{3}$ off a mattress–box springs combination. Find the sale price if the original price was $639.

$426.00

4–53. Angella bought two dresses on sale at 30% off and saved $41 from the marked price. What was the original price of the dresses?

$41 / .30 = $136.67

4–54. A grain dealer bought peanut hearts for $8.70 per bag and sold them for $11.60. Find his percent of profit.

33%

4–55. How long must $30,000 be invested at $7\frac{1}{2}\%$ to yield $6750?

3 years

4–56. Find the sale price of an outboard motor that was marked 22% off a list price of $589.

$459.42

4–57. How much would you save if you received a $3\frac{1}{2}\%$ discount for paying cash for a $595 TV set?

$20.83

4–58. A lumber company allows an 8% contractor discount. If a contractor bought lumber worth $8256, how much would the contractor pay for the lumber after the discount?

$7595.52

4–59. A hardware store sold a radial arm saw for $392.50. The store bought the saw for 65% of the selling price. What did the store pay for the saw?

$255.13

4–60. A real estate agent, Kay Cobb, bought two lots for $12,000 and $9000. She sold the first lot at a profit of 8% but lost 10% on the second. How much did she gain or lose on these deals?

Gained $60

Chapter 4 / Percentages

DISCOUNTS

Companies generally give discounts for volume buying. An additional discount is given for cash payment or payment within a short period of time. Finding the percent of increase or decrease, profit or loss, and the percent of change is necessary when solving applied problems.

Example 4–14

PROBLEM: Find the price paid for $240 of wool fabrics if discounts of 20% and 2% are given.

SOLUTION: Find the 20% discount by multiplying the original price ($240) by the rate of discount (0.20).

$$d = \$240 \times 0.20 = \$48$$

Subtract the first discount from the original price.

$$\begin{array}{rl} \$240.00 & \text{original price} \\ -\ 48.00 & \text{first discount} \\ \hline \$192.00 & \text{discount price} \end{array}$$

Find the 2% cash discount by multiplying the discount price ($192) by the rate of discount (0.02).

$$d = \$192 \times 0.02 = \$3.84$$

Then subtract the cash discount from the discount price.

$$\begin{array}{rl} \$192.00 & \text{discount price} \\ -\ 3.84 & \text{cash discount} \\ \hline \$188.16 & \text{price paid} \end{array}$$

ALTERNATIVE SOLUTION: Since the discounted price is 80% of the original price, subtract 20% from 100% and multiply 0.80 by the original price. Multiply the decimal 0.80 by the original price ($240) to determine the discounted price:

$$\text{discounted price} = \$240 \times 0.80 = \$192$$

Then subtract 2% from 100% and change to a decimal. Multiply the decimal 0.98 by the discounted price $192 to determine the price paid:

$$\text{price paid} = \$192 \times 0.98 = \$188.16$$

Example 4–15

PROBLEM: Find the percent of increase in yield of wheat by applying fertilizer, if the yield increases from 48 bushels to 64 bushels.

SOLUTION: Find the amound of change.

$$\begin{array}{rl} 64 & \text{bushels yield with fertilizer} \\ -48 & \text{bushel yield without fertilizer} \\ \hline 16 & \text{bushel increase} \end{array}$$

Divide the amount of change (16) by the yield without fertilizer (48) to find the percent of change:

$$\text{percent of change (increase)} = \frac{\text{change}}{\text{original}} = \frac{16}{48}$$

Note: $\frac{16}{48}$ should be reduced to $\frac{1}{3}$ or $33\frac{1}{3}\%$. Then change the decimal to a percent.

$$0.333 = 33\frac{1}{3}\%$$

Example 4–16

PROBLEM: A basketball player makes 10 out of 14 shots. What percent does he make?

SOLUTION: The percentage is 10 and the base is 14. The rate or percent must be found. To find the rate, use the formula

$$\text{rate} = \frac{\text{percentage}}{\text{base}} \quad \text{or} \quad r = \frac{P}{b}$$

Identify the given quantities of the formula.

$$r = \frac{P}{b} = \frac{10}{14}$$

Second, divide the percentage 10 by the base 14 to determine the rate.

$$r = \frac{10}{14} = 0.7142$$

Change the decimal to the nearest tenth of a percent.

$$0.7142 = 71.4\%$$

Thus the basketball player made 71.4% of his shots. That is, his shooting average in basketball is 71.4%.

Example 4–4

PROBLEM: Solve the following applied problems.

4–61. A soap is advertised as $99\frac{44}{100}$% pure. How many grams of impurities would there be in 5000 grams?

4912

28 grams

4–62. A service charge of $1\frac{1}{2}$% is charged on the unpaid balance of charge accounts. Determine the interest on a bill for $59.95.

4–63. A real estate salesperson receives a 3.4% commission on the sale of a house. How much commission would she receive for selling a $144,500 house?

$4913.00

4–64. A trucking firm charges 3% commission for hauling and collects another $1\frac{1}{3}$% for insurance. What are the fees for a $12,000 cargo?

$164.8   360
*   160*
* $520.00*

4–65. A certain crude oil is 14.875% sulfur. How many gallons of sulfur would there be in 8000 barrels of crude oil if each barrel contained 31.5 gallons?

1190
37,485 gallons

4–66. If 37.42% is a "factor" for hot-dog buyers at baseball games, determine the profit on hot dogs for the season. Eighty-one games are played and the average attendance is 24,276. Hot dogs sell for $1.25 each and the profit on the hot-dog sales is 62%.

$570,253.07 $.775

4–67. At a baseball game, 37.42% of the people bought hot dogs. How much money was collected if 27,753 people attended the game and the hot dogs sold for $1.25?

10 385.1726
= $12,981.47

4–68. A tolerance of ±0.8% is allowable for a car axle. How much variation is allowable on $\frac{7}{8}$-inch axle?

T .007 in.   .875
.882 Between .868   44/500   217/250

4–69. What are the upper and lower limits of tolerance of a 910-ohm resistor if the tolerance factor is ±0.5%?

4.55

upper Tolerance = 914.55 Ω
Lower = 905.45 Ω

4–70. Find the efficiency of an electric motor that draws 54 watts from an electrical supply and delivers 46.6 watts to its load.

86.3%

4–71. Minnesota iron ore is 18.1% pure. How much iron can be refined from 150 tons of iron ore?

27.15 TONS

4–72. The list price of an office chair is $229.89. The interior decorator's discount is 40%. What does the decorator pay for the chair if the state sales tax is 4%?

137.93
5.52
$143.45

4–73. A 2-meter length of nylon rope was given a tensile test and stretched to 2.23 meters before breaking. Determine the percent of the original length it was stretched.

11.5%

4–74. A drill-press operator was to drill a hole 0.3125 inch in diameter. The hole drilled measured 0.34375. Find the percent of error.

.34375
−.3125
.03125 / .3125 = 10%

4–75. Albert bought a lakeshore cabin for $23,600 and later sold it for 25% more than he paid for it. The realtor charged a 6% fee on the selling price. What profit did Albert make?

29500
× .96%
28320

2774.
−23
1770

$4130

4–76. A lawn fertilizer is 22% nitrogen. If a 50-pound bag costs $2.98, find the price per pound for nitrogen.

$2.98 = $.27
50×.22= 11
50 × .22%

4–77. If $1\frac{1}{4}$ pounds of salt are dissolved in 5.5 gallons of water that weigh 45.8 pounds, what is the percent of salt in the solution to the nearest hundredth?

2.73%

4–78. How much would you need to invest at 7.5% if you wanted an income of $1200 per month?

$16,000   14400
19200

4–79. The cost of living rose 6.3%, 8.1%, and 5.8% each year over a 3-year period. The raise in pay amounted to 25% over 3 years. How much is a person ahead or behind at the end of 3 years if his initial yearly salary is $20,500?

.202 + $984.00

25625.
24922.89
$702.11 ahead.

4–80. The depreciation of a new car is approximately 18% of the original price the first year, 12.5% of the original price the second year, and 8% of the original price the third year. The $12,900 car loan is at 13.5% per year for 3 years. Determine the depreciation and interest for 3 years.

Year
1) $10,578
2) $9255.75
3) $8545.29

= 4966.5
= 5224.50

Total Depreciation = $4384.71

THINK TIME

Many people have difficulty understanding percents. A free-spirited instructor, Scotty Jacobs, developed a system to translate percent language to simple arithmetic. The system can best be explained by examples.

PROBLEM: What is 20% of $250?

JACOB'S LAW: ? = 0.20 × 250. Thus "what" becomes "?"; "is" becomes "="; "20%" becomes "0.20"; and "of" becomes "times" or "×."

$$? = 0.20 \times \$250.$$
$$? = \$50$$

PROBLEM: 52 is what percent of 260?

JACOB'S LAW: 52 = ? × 260. Thus "is" becomes "=," "what" becomes "?", and "of" becomes "×."

$$52 = ? = 260$$
$$\frac{52}{260} = \frac{? \times 260}{260}$$
$$\frac{52}{260} = ?$$
$$0.20 = ?$$
$$20\% = ?$$

Check: 52 = 20% of 260
           = 0.20 × 260
           = 52

PROBLEM: 105 is 75% of what number?

JACOB'S LAW: 105 = 0.75 × ?. Thus "is" becomes "=," "of" becomes "×,"' and "what" becomes "?".

To solve, divide both sides by .75.

$$105 = 0.75 \times ?$$
$$\frac{105}{0.75} = \frac{0.75 \times ?}{0.75}$$
$$\frac{105}{0.75} = ?$$
$$140 = ?$$

Check: 105 = 0.75 × 140
            = 105

PROCEDURES TO REMEMBER

1. To change a percent to a decimal, remove the percent sign and move the decimal point two places to the left.
2. To change a decimal to a percent, move the decimal point two places to the right and add the percent sign to the right of the number.
3. To change a fraction to a percent, divide the bottom number (denominator) into the top number (numerator) of the fraction; then move the decimal point two places to the right, and add the percent sign to the right of the number.
4. To change a percent to a common fraction, remove the percent sign, write the number over 100, and reduce to lowest terms.
5. To find the percent of a number, convert the percent to a decimal and multiply.
6. To find the selling price of an article based on a percentage of profit:

   Method 1
   (a) Change the percent to a decimal.
   (b) Multiply the cost by the decimal to obtain the profit.
   (c) Add the profit to the cost to determine the selling price.

   Method 2
   (a) Consider the cost as 100%.
   (b) Add the percent of profit and the percent of cost.
   (c) Multiply the cost by the sum of percents to determine the selling price.

7. To determine the discount from the selling price:

   Method 1
   (a) Change the percent of discount to a decimal.
   (b) Multiply the selling price by the decimal to find the discount.
   (c) Subtract the discount from the selling price to determine the discounted price.

   Method 2
   (a) Consider the selling price as 100%.
   (b) Subtract the percent of discount from 100%.
   (c) Subtract the discount from the selling price to determine the discounted price.

8. To find interest, multiply the rate *times* the principal *times* the amount of time the loan is in effect ($i = prt$).
9. To find the base or principal, convert the percent or rate of interest to a decimal and divide the interest by the rate of interest in decimal form.
10. To determine the rate of interest, divide the interest by the base or principal and convert decimal to percent.

CHAPTER SUMMARY

1. *Percent* means part of 100 or the comparison of one number to another number.
2. The symbol for percent is %.
3. There are three types of percentage problems:
   (a) What is 5% of 265?
   (b) Sixteen is what percent of 64?
   (c) Thirty-nine is 20% of what number?
4. Percent problems involve three quantities:
   (a) The base.
   (b) The rate.
   (c) The percentage.
5. Percent formulas
   (a) *Percentage* equals base *times* rate ($p = b \times r$).
   (b) *Base* equals percentage *divided* by rate $\left(b = \dfrac{P}{r}\right)$.
   (c) *Rate* equals percentage *divided* by base $\left(r = \dfrac{P}{b}\right)$.
6. *Interest* is money paid for the use of money and is calculated by a percentage of the money borrowed.
7. *Rate* of interest is the percent of interest per year.
8. *Base* is the principal or amount of money loaned.
9. *Interest* equals principal *times* rate *times* time ($i = prt$).
10. *Principal* equals interest *divided* by rate *times* time $\left(p = \dfrac{i}{rt}\right)$.
11. *Rate of interest* equals interest *divided* by principal *times* time $\left(r = \dfrac{i}{pt}\right)$.

CHAPTER TEST

T–4–1. Change $\dfrac{7}{8}$ to a decimal.

.875

T–4–2. Change $14\dfrac{2}{5}$ to a decimal.

14.4

T–4–3. Convert 45% to a common fraction.

9/20

T–4–4. Convert $37\dfrac{1}{2}$% to a common fraction.

3/8

T–4–5. Convert 0.31 to a percent.

31%

T–4–6. Convert 1.25 to a percent.

125%

T–4–7. Convert 0.5% to a decimal.

.005

T–4–8. Write $\dfrac{1}{4}$% as a decimal.

.0025

T–4–9. Convert 0.32% to a common fraction.

8/625

T–4–10. Convert $7\frac{3}{4}$% to a decimal.

.0775

T–4–11. $78\% = \dfrac{?}{100}$

78

T–4–12. Convert $\dfrac{7.5}{10}$ to a percent.

75%

T–4–13. $\dfrac{3}{33\frac{1}{3}} = \dfrac{?}{100}$

x = 9

T–4–14. Find 25% of 78.9.

19.725

T–4–15. Find $7\frac{1}{4}$% of 5764.

417.89

T–4–16. A family sold a lot for $20,000 and gained 25% on the transaction. What did the lot cost the family?

16000

T–4–17. What is the percent of increase from the application of fertilizer on a wheat field whose yield was 24 bushels before and 38 bushels after?

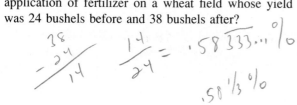

.58 1/3 %

T–4–18. A man bought a house for $82,000 and 5 years later sold it for $102,500. Find the percent of gain.

25%

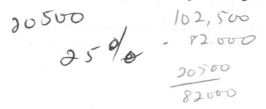

T–4–19. An automobile depreciated 24% and was worth $8800 at the end of the year. What was it worth at the beginning of the year?

8800 / .76 = 11578.94737

T–4–20. A lathe is set to operate at 4200 revolutions per minute. The belt slippage is 18%. What is the actual speed of the machine?

3444 Rev per min.

T-4-21. Brucie works at $7.32 per hour for 22 hours per week. Find her take-home pay if 27.2% is deducted for taxes, social security, and other benefits.

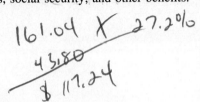

T-4-22. A set of golf clubs originally priced at $475 was discounted at 40%, 30%, and 10%. What was the final sale price?

475 × .6 × .7 × .9 = $179.55

T-4-23. The retail price of a radial tire is $105. If the tire store is allowed a 25% profit and the manufacturer earns a 20% profit, what is the cost of making the tire?

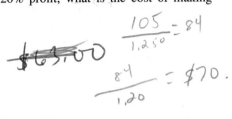

$\frac{105}{1.250} = 84$

$\frac{84}{1.20} = \$70.$

T-4-24. How much money would you need to invest at $7\frac{1}{2}\%$ to have an income of $750 per month?

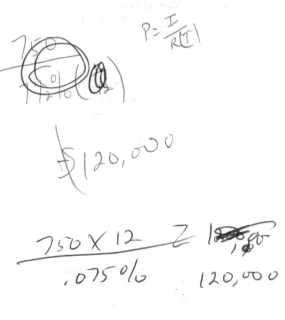

T-4-25. A plastic material loses 4.25% of its weight in drying. What was its original weight if its weight after drying is 134.05 grams?

$100 - 4.25\% = .9575$

$\frac{134.05}{.9575}$

140.00

5

POWERS AND ROOTS

OBJECTIVES

1. To learn and review powers of 10, scientific notation, and their application.
2. To learn and review the skill to solve square roots.
3. To develop the skills to use powers and roots to solve applied problems.

SELF-TEST

This test will determine your need to review powers of 10 and square roots. Show your work; this will indicate areas where you may need to improve.

S–5–1. Write 5000 as a power of 10.

5 × 10³

S–5–2. Write 589,000 as a power of 10.

5.89 × 10⁵

S–5–3. Convert 7×10^5 to a number.

700,000

S–5–4. Convert 2.4×10^6 to a number.

2,400,000

S–5–5. Convert 0.0000052×10^8 to a number.

S–5–6. Write 0.0002 as a power of 10.

2 × 10⁻⁴

S–5–7. Write 0.000000789 as a power of 10.

7.89 × 10⁻⁷

S–5–8. Convert 4.54×10^{-1} to a number.

.454

67

S–5–9. Convert 779×10^{-4} to a number.

.0779

S–5–10. Convert $0.000000000126 \times 10^{12}$.

126.

S–5–11. $21^2 =$ 441

S–5–12. $0.00494^2 =$

2.4404×10^{-5}
.000024404

S–5–13. $6^3 =$ 216

S–5–14. $\left(\dfrac{3}{8}\right)^3 = \dfrac{27}{512}$

S–5–15. $3^6 =$ 729

S–5–16. Calculate the square root of 141,000 to the nearest tenth.

375.5

S–5–17. Evaluate 6.35^3.

256.047875

S–5–18. Calculate the square root of 0.120409.

.347

S–5–19. Calculate the square root of 24,409 to the nearest thousandth.

156.234

S–5–20. Calculate the square root of 5280 to the nearest hundredth.

72.664

S–5–21. Find the diagonal of a square with 15-foot sides to the nearest foot.

21.2 ft

S–5–22. Find the hypotenuse of a right triangle if one leg is 12 units and the other leg is 5 units.

13 units

S–5–23. Find the cube root of 343.

7

S–5–24. Find the length, to the nearest tenth, of the side of a square whose diagonal is 16.97 meters.

12 meters

S–5–25. Find the side of a square, to the nearest tenth, if the diagonal is 18.5.

13.1

INTRODUCTION

The power-of-10 concept was discussed in Chapter 3. In this chapter we develop the power-of-10 concepts further and learn scientific notation and how to determine roots. Knowledge of powers of 10 helps estimate answers when working with very large or very small numbers. The hand-held calculator is the most efficient way to work with powers of 10. However, it is necessary to have an understanding of how to convert numbers to powers or to convert powers-of-10 quantities to numbers.

DEFINITIONS

When working with very large or very small numbers it is helpful to convert numbers to a *power of 10*. *Scientific notation* is the procedure of rewriting a number as a number between 1 and 10 times the appropriate power of 10. A number is said to be in scientific notation if it is expressed as the product of a single digit number and some power of 10. For example, 4,000,000 could be written $4 \times 1,000,000 = 4 \times 10^6$. The decimal point of any number can be shifted to the left or right depending on the power of 10. For example, in $5.25 \times 10^3 = 5200$ the 3 adjacent to the 10 is read "10 to the third power" or "10 to the third," which means to move the decimal point three places to the right. Also, $7.46 \times 10^{-2} = 0.0746$; the -2, read "10 to the negative 2 power," means to move the decimal point two places to the left.

Engineering notation is rewriting a number so as to use the powers of 10, where an appropriate SI prefix is attached (see Chapter 18). For example, 7.5 kW means 7.5×10^3 W or 7500 watts, and 8 microseconds = 0.000008 second = 8×10^{-6} second, which is written as 8 μs in engineering notation. This notation is used in technical fields, particularly in electronics. Figure 5–1 shows powers of 10. Note the power of 10 is the same digit as the number of zeros in the original numbers.

Note the negative power of 10 is the number of places the decimal would need to be moved to the right so that the first digit would be more than 1.

A number is written in scientific notation when the first factor is a single digit and the second is a power of 10. Thus

$$5000 = 5 \times 1000 = 5 \times 10^3$$

$$63,700 = 6.37 \times 10,000 = 6.37 \times 10^4$$

$$0.007 = 7 \times 0.001 = 7 \times 10^{-3}$$

Greater Than 1	Less Than 1
$1 = 10^0$	
$10 = 10^1$	$10^{-1} = \frac{1}{10^1} = \frac{1}{10} = 0.1$
$100 = 10^2$	$10^{-2} = \frac{1}{10^2} = \frac{1}{100} = 0.01$
$1000 = 10^3$	$10^{-3} = \frac{1}{10^3} = \frac{1}{1000} = 0.001$
$10,000 = 10^4$	$10^{-4} = \frac{1}{10^4} = \frac{1}{10,000} = 0.0001$
$100,000 = 10^5$	$10^{-5} = \frac{1}{10^5} = \frac{1}{100,000} = 0.00001$
$1,000,000 = 10^6$	$10^{-6} = \frac{1}{10^6} = \frac{1}{1,000,000} = 0.000001$

Figure 5-1

POWERS OF 10

The following examples show how to convert numbers to powers of 10 and how to convert power-of-10 quantities to numbers.

Example 5-1

PROBLEM: Write 9000 as a power of 10.

SOLUTION: $9000 = 9 \times 1000 = 9 \times 10^3$

The power of 10 is equal to the number of zeros in the original number.

Example 5-2

PROBLEM: Write the value of 5.23×10^5.

SOLUTION:
$5.23 \times 10^5 = 5.23 \times 100,000 = 523,000$

The number of zeros is equal to the number of places moved to the right or the power of 10.

Example 5-3

PROBLEM: Write 0.000008 as a power of 10.

SOLUTION:
$0.000008 = 8 \times 0.000001 = 8 \times 10^{-6}$

The negative 6 indicates the number of places the decimal needs to be moved to the left.

Example 5-4

PROBLEM: Write the value of 5.69×10^{-4} as a number.

SOLUTION:
$5.69 \times 10^{-4} = 5.69 \times 0.0001 = 0.000569$

The negative 4 indicates how many places to the left the decimal point should be moved.

EXERCISE 5-1

Write each of the following as a power of 10.

5-1. 3000

3×10^3

5-2. 150

1.5×10^2

5-3. 46,000

4.6×10^4

5-4. 567,000

5.67×10^5

5-5. 8,760,000,000

8.76×10^9

Convert the following to numbers.

5–6. 4×10^3

4000

5–7. 3.9×10^2

390

5–8. 6.66×10^5

666,000

5–9. 0.000036×10^6

36

5–10. 1.234×10^4

12,340

Write each of the following as a power of 10.

5–11. 0.007

7×10^{-3}

5–12. 0.00086

8.6×10^{-4}

5–13. 0.29

2.9×10^{-1}

5–14. 0.000445

4.45×10^{-4}

5–15. 0.00000000053

5.3×10^{-10}

Convert the following to numbers.

5–16. 4.43×10^{-2}

.0443

5–17. 0.0038×10^2

0.38

5–18. 4.1000×10^{-2}

0.041

5–19. 250×10^{-3}

0.25

5–20. $0.00000000764 \times 10^{12}$

7640.0

SQUARE ROOT

The calculation of square root is essential to the solution of many problems. There are several long ways to calculate square roots; however, the easiest and most accurate is with a calculator. To find the square root of a number means to determine what number, when multiplied by itself, will equal the original number. For example, the square root of 16 is 4, since 4 × 4 equals the original number, 16. Few numbers have simple square roots. For example, the square root of 5 is not an exact number—it can only be approximated. The square root of 5, written as $\sqrt{5}$, is known as an irrational number because it will not end but will continue as long as the process is continued. The easiest way to calculate square root is with an electronic calculator. Another way is to use a table of square roots; however, the use of a calculator is recommended.

To find the *square root* of a number means to find a number which, when multiplied by itself, will equal the original number. When several numbers are multiplied, the numbers being multiplied are called *factors*. In the example 5 × 3 × 2 = 30, the 5, 3, and 2 are factors. When a number is used as a factor more than once, it may be written with an exponent. An *exponent* is a number written above and to the right of another number and indicates how many times that number is used as a factor. For example, $5^2 = 5 \times 5 = 25$; the 2 is an exponent which indicates that the number 5 is used as a factor two times. When a factor is used two times, it is known as being *squared*. A number is *cubed* when it is used as a factor three times. For example, $2^3 = 2 \times 2 \times 2 = 8$—we say that the factor 2 is cubed. When a factor is used more than three times, it is generally called a *power*. For example, $3^4 = 3 \times 3 \times 3 \times 3 = 81$—the 4 indicates that the 3 is used as a factor four times. This can also be read as "3 to the fourth power."

The square root of a number is a number that when multiplied by itself is equal to the original number. The square-root symbol $\sqrt{}$ indicates to find a number that when multiplied by itself will equal the number inside the symbol. Thus $\sqrt{49} = 7$ since 7×7 *or* $7^2 = 49$. Similarly, the cube root of a number is one of three equal factors, and the fourth root of a number is one of four equal factors.

The *radical symbol* $\sqrt[3]{8}$ indicates the third root of 8 is to be found. The 3 inside the radical symbol is called the *index of the root* and indicates what root is to be determined. When no index appears in the radical symbol the number 2 or square root is understood. The number under the radical symbol, 8, is called the *radicand*. Numbers that have even square roots are called perfect squares. Examples of *perfect squares* are 1, 4, 9, 16, 25, 36, and so on.

EXPONENTS AND FACTORS

The following examples illustrate the use of exponents and factors. Refer to the definitions given to recall the meanings of the terms.

Example 5–5

PROBLEM: Write 3 × 3 × 3 × 3 × 3 in an abbreviated form and evaluate.

SOLUTION: Since 3 is used as a factor five times, it is written as 3^5. Then multiply

$$3 \times 3 \times 3 \times 3 \times 3 = 243$$
$$3^5 = 243$$

Or use the $\boxed{x^y}$ key on your calculator as follows: Press $\boxed{3}$, then press the $\boxed{x^y}$ key, press $\boxed{5}$, and then $\boxed{=}$; 243 should appear in the display.

Example 5–6

PROBLEM: Write 5 × 5 × 2 × 2 × 2 in an abbreviated form and evaluate.

SOLUTION: Since 5 is used as a factor two times, it is written as 5^2; 2 is used as a factor three times, and is written as 2^3. Multiply as follows:

$$5 \times 5 \times 2 \times 2 \times 2 =$$
$$25 \times 8 = 200$$

A calculator should be used to solve this problem.

Example 5–7

PROBLEM: What does fifteen squared equal?

SOLUTION: Fifteen squared, or 15^2, means to multiply 15 × 15, which equals 225. Use the $\boxed{x^2}$ key on your calculator as follows:

$$\boxed{15}\boxed{x^2} = 225$$

Example 5–8

PROBLEM: What does eight cubed equal?

SOLUTION: Eight is used as a factor three times, which can be written as 8^3.

$$8^3 = 8 \times 8 \times 8 = 512$$

To use your calculator, press $\boxed{8}$, then $\boxed{x^y}$, then $\boxed{3}$, then $\boxed{=}$.

EXERCISE 5–2

Solve the following problems to practice working with factors and exponents. Use a calculator and the procedures shown in the examples.

5–21. Evaluate 5^3.

125

5–22. Evaluate $4^2 \times 4^2$.

256

5–23. Evaluate 4^5.

1024
~~*100,000*~~

5–24. Evaluate 12^2.

144

5–25. Evaluate 150^2.

22,500

5–26. Write $2 \times 2 \times 2 \times 4 \times 4$ using exponents, and then evaluate.

$2^3 \times 4^2 = 128$

5–27. Write $3 \times 3 \times 3 \times 3 \times 3 \times 3$ using exponents, and then evaluate.

$3^6 = 729$

5–28. Evaluate 12^3.

1728

5–29. Evaluate $\left(\dfrac{1}{2}\right)^4$.

.0625 or 1/16

5–30. Evaluate $\dfrac{1}{2} \times \dfrac{1}{2} \times \dfrac{1}{2} \times 4 \times 4 \times 2 \times 2 \times 2$.

$\left(\dfrac{1}{2}\right)^3 \times (4)^2 \times (2)^3 = 16$

5–31. Evaluate 0.5^2.

.25

5–32. Evaluate 0.22^2.

.0484

$(.22)^3 = .010648$

Chapter 5 / Powers and Roots

5–33. Evaluate 0.8^3.

.512

5–34. Evaluate 0.102^2.

.010404

5–35. Evaluate 0.12^3.

.001728

5–36. Evaluate 3.15^2.

9.9225

5–37. Evaluate 10.1^3.

1030.301

5–38. Evaluate 0.4^4.

.0256

5–39. Evaluate 2.5^5.

97.65625

5–40. Evaluate $8^2 \times 8^2$.

4,094

SQUARE ROOT CALCULATION

As you observe the longhand method of calculating square root as shown in the example, keep in mind this is one of the methods that may be used. You should also note that most square-root calculations determine only an approximate value. The extent of accuracy will depend on the individual problems. This example is included only to illustrate the difficulty of calculating square root. As the numbers become more complex, the calculation becomes more difficult.

Example 5–9

PROBLEM: Calculate the square root of 576.

SOLUTION: Set up the square-root process. Place 576 under the radical symbol. Divide the digits into groups of two, starting from the decimal point and moving to the left.

$$\sqrt{5_\wedge 76.}$$

From the first group, 5 and 2, find the largest perfect square smaller than 5. This is 2, since $2^2 = 4$. Place the 2 above the number 5 and the number 4 below the number 5. Subtract the number 4 from the number 5 and bring down the next group, 76.

$$\begin{array}{r} 2 \\ \sqrt{5_\wedge 76.} \\ 4 \\ \hline 1\ 76 \end{array}$$

Multiply 20 times 2, the number above the radical sign. Then divide the result of 20×2, or 40, into 176. This will give you the approximate quotient, which is 4. Place 4 above the number 6 on top of the radical sign.

$$\begin{array}{r} 4 \\ 40\overline{)176.00} \\ 160 \end{array} \qquad \begin{array}{r} 2\ 4 \\ \sqrt{5_\wedge 76} \\ 4 \\ \hline 1\ 76 \end{array}$$

Then add the approximate quotient 4 to 40, $4 + 40 = 44$. Then multiply 44×4 and place the result below the 176. Subtract the resulting number from 176.

```
                    2 4
                  √5₍76
   4      4      4
 +40     ×40    ─────
 ───    ───     1 76
  44    176     1 76
                ─────
                0 00
```

Since the resulting subtraction of 176 from 176 equals zero, we have a perfect square of 24. Check your answer by multiplying 24 by 24.

```
   24
  ×24
  ───
   96
   48
  ───
  576
```

Once you understand the procedure, the process can be condensed as follows:

```
                            20
 √5₍76      √5₍76.         × 2
             4             ───
            ───             40
            1 76
```

```
    4        40        2 4
 40)176     + 4      √5₍76
    160     ───       4
            44        ───
            × 4       1 76
            ───       1 76
            176       ─────
                      0 00
```

```
CHECK:       24
            ×24
            ───
             96
             48
            ───
            576
```

SQUARE-ROOT APPLICATIONS

Square-root skills are necessary to determine unknown distances. For example, carpenters use square-root skills to find the lengths of rafters. The formula for finding the length of a rafter is based on the following formula: The square of the hypotenuse of a right triangle is equal to the square of one side plus the square of the other side. Study Figure 5–2. The hypotenuse is equal to the

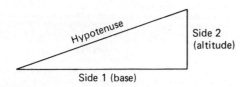

Figure 5–2

square root of side 1 squared plus side 2 squared. Study the following formulas.

$$\text{Hypotenuse} = \sqrt{\text{base}^2 + \text{altitude}^2}$$
$$\text{Base} = \sqrt{\text{hypotenuse}^2 - \text{altitude}^2}$$
$$\text{Altitude} = \sqrt{\text{hypotenuse}^2 - \text{base}^2}$$
$$\text{Diagonal side} = \sqrt{(\text{diagonal side})^2 - (\text{side 2})^2}$$
$$\text{Side 1} = \sqrt{(\text{diagonal side})^2 - (\text{side 2})^2}$$
$$\text{Side 2} = \sqrt{(\text{diagonal side})^2 - (\text{side 1})^2}$$

These formulas are valid only for *right* triangles.

Example 5–10

PROBLEM: Find the length of a rafter to the nearest $\frac{1}{8}$ inch if the rise is 6 feet and the run is 12 feet. See Figure 5–3.

Figure 5–3

SOLUTION: To find the length of a rafter, take the square root of the square of the rise (6 feet) plus the square of the run (12 feet). Use the formulas shown.

$$\text{Rafter length} = \sqrt{(\text{rise})^2 + (\text{run})^2}$$
$$= \sqrt{6^2 + 12^2}$$
$$= \sqrt{36 + 144}$$
$$= \sqrt{180}$$

Using a calculator, key in 180 and press $\boxed{\sqrt{\ }}$. The display will then show 13.416408, which means 13.416408 feet. To convert to the nearest $\frac{1}{8}$ inch, from 13.416408 feet, subtract 13 feet.

$$13.416408 - 13 = 0.416408 \text{ ft}$$

Then multiply 0.416408 by 12 inches. (12 inches = 1 foot.)

$$0.416408 \text{ ft} \times 12 \text{ in./ft} = 4.9968943 \text{ in.}$$

From 4.9968943 inches, subtract 4 inches.

$$4.9968943 \text{ in.} - 4 \text{ in.} = 0.9968943 \text{ in.}$$

Chapter 5 / Powers and Roots

Multiply 0.9967943 inch by $\frac{8}{8}$ to convert the decimal part of an inch to eighths.

$$0.9968943 \times \frac{8}{8} = \frac{7.9751546}{8}$$

This rounds to $\frac{8}{8}$ inch or 1 inch; thus the 4.9968943 becomes 5 inches to the nearest $\frac{1}{8}$ inch. Thus the rafter would be 13 feet 5 inches.

EXERCISE 5-3

Use a calculator to solve the folllowing. Where appropriate, calculate the roots to the same number of digits as in the original number.

5-41. $\sqrt{121}$

11

5-42. $\sqrt{256}$

16

5-43. $\sqrt{1024}$

32

5-44. $\sqrt{3136}$

56

5-45. $\sqrt{5476}$

74

5-46. $\sqrt{708,964}$

842

5-47. $\sqrt{0.0576}$

0.24

5-48. $\sqrt{38.4400}$

6.20

Calculate the square root to the nearest thousandth.

5-49. $\sqrt{0.6561}$

.810

5-50. $\sqrt{0.002116}$

0.046

5-51. $\sqrt{38}$

6.164

5-52. $\sqrt{149}$

12.207

5-53. $\sqrt{85}$

9.220

5-54. $\sqrt{5280}$

72.664

5–55. $\sqrt{0.1234}$

0.351

5–56. $\sqrt{10,000}$

100

5–57. $\sqrt{30.603024}$

5.532

5–58. $\sqrt{0.60516}$

.778

5–59. $\sqrt{1989}$

44.6
44.598

5–60. $\sqrt{91.550451}$

9.568

5–61. Find the length of the diagonal (hypotenuse) of a 24-meter square to the nearest tenth.

$\sqrt{24^2 + 24^2}$ 33.9 m

5–62. Find the length of the diagonal (hypotenuse) of a 14.5-centimeter square to the nearest hundredth.

$\sqrt{14.5 + 14.5}$ = 20.51 cm

5–63. Find the length of the diagonal of the rectangle in Figure 5–4 to the nearest tenth.

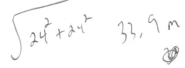

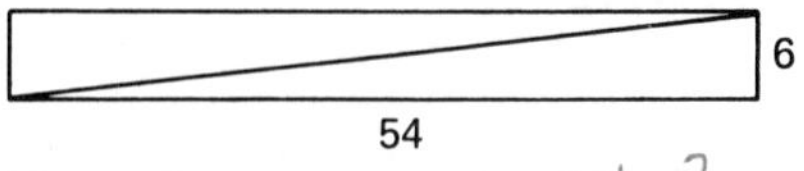

Figure 5–4

54.3

5–64. Find the altitude of the triangle in Figure 5–5 to the nearest tenth.

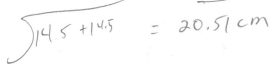

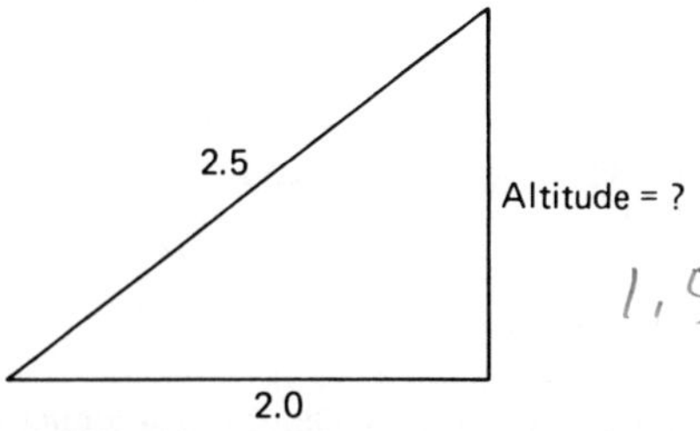

Figure 5–5

1.5

5–65. Find the hypotenuse of the triangle in Figure 5–6 to the nearest thousandth.

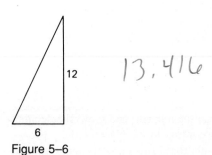

13.416

Figure 5–6

5–66. Find the altitude of a right triangle whose hypotenuse is 20 and whose base is 16.

(12)

$\sqrt{20^2 - 16^2}$

Chapter 5 / Powers and Roots

5–67. Find the hypotenuse of the triangle in Figure 5–7 to the nearest tenth.

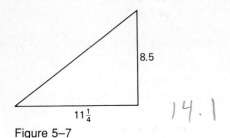

Figure 5–7

5–68. Find the rise if the rafter is 37.1 feet and the run is 36 feet to the nearest tenth of a foot.

5–69. Find the length of the rafter for the roof in Figure 5–8 to the nearest hundredth.

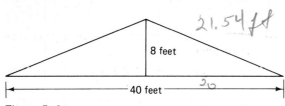

Figure 5–8

5–70. Good proportioning in printing requires the diagonal be two times the width of the page. To the nearest hundredth, find the length of the page if the width is 24 centimeters.

$\sqrt{48^2 - 24^2} = \sqrt{1728}$

41.57 cm

THINK TIME

Many different methods can be used to approximate the square root of a number. Study the approximation procedure shown. A basic function calculator will be very helpful in finding the approximation.

PROBLEM: Find the approximate square root of 58.

SOLUTION: Recall the perfect squares smaller and larger than 58.

$$7 \times 7 = 49 \quad \text{and} \quad 8 \times 8 = 64$$

Make 7 the first approximation, $A_1 = 7$. Then divide 58 by 7, and add 7; then find the average as shown. This result will give the second approximation, A_2.

$$A_2 = \frac{\frac{58}{7} + 7}{2} = \frac{8.2857143 + 7}{2}$$
$$= \frac{15.285714}{2} = 7.6428571$$

Now divide 58 by A_2, the second approximation, add A_2, and find their average. This will give the third approximation A_3.

$$A_3 = \frac{\frac{58}{7.6428571} + 7.6428571}{2}$$
$$= \frac{15.2311642}{2} = 7.6158244$$

Note: $(7.6158211)^2 = 58.000731$; thus the approximation is accurate to the nearest thousandth. If your first approximation is near the correct value, the second approximation will be reasonably accurate. This procedure may be used to find square roots.

PROCEDURES TO REMEMBER

1. To determine the scientific notation:
   (a) To convert a number to scientific notation, write the first digit of the number, place a decimal point, add the digit that follows, and then write the power of 10 according to the number of places the decimal point needs to be moved to achieve the original number. *Example:* $52,000 = 5.2 \times 10^4$.
   (b) To convert a power-of-10 quantity to a number, write the number, and move the decimal point according to the place of the power—to the right for positive powers and to the left for negative powers. *Examples:* $6.53 \times 10^4 = 65,300$; $7.84 \times 10^{-3} = 0.00784$.

2. To calculate the square root of a number:
   (a) Place the number under the radical symbol.
   (b) Divide the digits of the number into groups of two, starting from the decimal point and moving in both directions as necessary.
   (c) Begin with the group on the far left, find the largest perfect square that is equal to or less than the number in the group.
   (d) Place this root number above the group and above the radical symbol.
   (e) Multiply this root number by itself and place the result below the first group on the left.
   (f) Subtract, and bring down the next group to the right of the first group.
   (g) Multiply 20 times the number above the radical symbol.
   (h) Divide this approximate number into the dividend created in step (f). This will give you an approximate quotient.
   (i) Place this number above the second group above the radical symbol.
   (j) Multiply this group by the last digit of the number, and enter it under the number created in step (f).
   (k) Repeat steps (f) through (j) until the square root has been found or satisfactorily approximated by adding zeros to form groups of two as necessary to the right of the numbers to the right of the decimal point.

CHAPTER SUMMARY

1. *Scientific notation* is the procedure of rewriting a number as a single-digit number, placing the decimal point, adding the remaining digits (generally limited to three digits), and indicating the appropriate power of 10—right for the positive powers and left for negative powers.
2. To calculate the *square root* of a number means to determine what number, multiplied by itself, will equal the original number.
3. *Factors* are numbers that are multiplied together to form another number.
4. An *exponent* is a number written above and to the right of another number and indicates how many times that number is used as a factor.
5. To *square* a number means to multiply the number by itself.
6. To *cube* a number means that the number is used as a factor three times.
7. When a factor is used three or more times, it is called a *power*.
8. The *radical* symbol, $\sqrt[3]{78}$, indicates that the cube root is to be determined.
9. The *index of the root*, $\sqrt[3]{}$, is located inside the radical sign.
10. When no index number is given, square root is understood.
11. The number under the radical, $\sqrt{78}$, is called the *radicand*.
12. Numbers that have even square roots are called *perfect squares*.
13. The square of a hypotenuse of a right triangle is equal to the square of the base plus the square of the altitude.
14. The hypotenuse of a right triangle equals the square root of the square of the base plus the square of the altitude.
15. The base of the right triangle is equal to the square root of the square of the hypotenuse minus the square of the altitude.
16. The altitude of a right triangle is equal to the square root of the hypotenuse minus the square of the base.

CHAPTER TEST

T–5–1. Write 9000 as a power of 10.

9×10^3

T–5–2. Write 789,000,000 as a power of 10.

7.89×10^8

T–5–3. Convert 8×10^6 to a number.

8,000,000

T–5–4. Convert 4.2×10^5 to a number.

420,000

T–5–5. Convert 0.000000056×10^9 to a number.

56

T–5–6. Write 0.005 as a power of 10.

5.0×10^{-3}

T–5–7. Write 0.00000648 as a power of 10.

6.48×10^{-6}

T–5–8. Convert 5.63×10^{-2} to a number.

0.0563

T–5–9. Convert 357×10^{-3} to a number.

0.357

T–5–10. Convert $0.0000000000000789 \times 10^{15}$ to a number.

78.9

T–5–11. $(23.2)^2 =$

538.24

T–5–12. $(0.949)^2 =$

0.900601

T–5–13. $(7)^3 =$

343

T–5–14. Find the square of 1.4142 to the nearest hundredth.

1.19

T–5–15. $(4)^4$

256

T–5–16. Calculate the square root of 141,376.

376

T–5–17. $(5.36)^3 =$

153.990656

T–5–18. Calculate the square root of 0.22373 to the nearest thousandth.

0.473

T–5–19. To the nearest hundredth, calculate the square root of 1609.

40.11

T–5–20. Calculate the square root of 26,606 to the nearest thousandth.

163.113

T–5–21. Find the diagonal of an 8-inch by 10-inch rectangle to the nearest tenth.

12.8 Inch

T–5–22. Find the hypotenuse of a right triangle if one side is 12.5 units and the other side is 16.4 units (to the nearest tenth).

20.6 units.

T–5–23. Find the cube root of 216.

6

T–5–24. Find the length of the side of a square if the diagonal is 10.6066 (to the nearest tenth).

7.5

T–5–25. What is the longest line that could be drawn on a basketball court that is 88 feet by 50 feet to the nearest hundredth foot?

101.21 ft

section two: basic algebra

6

DEFINITIONS AND BASIC OPERATIONS OF ALGEBRA

OBJECTIVES

1. To define algebraic terms.
2. To perform basic algebraic operatons.
3. To convert written problems to formulas or algebraic equations.

SELF-TEST

The test will help you evaluate your basic algebric skills. Show your work so that any difficulties can be noted and help obtained.

S–6–1. Subtract: $\quad -11xy$
$\quad\quad\quad\quad\quad\quad\quad +\ 7xy$

S–6–2. Subtract: $\quad +764$
$\quad\quad\quad\quad\quad\quad\quad -231$

S–6–3. Add: $\quad +42$
$\quad\quad\quad\quad\quad -26$

S–6–4. Add: $\quad -17$
$\quad\quad\quad\quad\quad +\ 6$

S–6–5. Divide: $\dfrac{-144}{-3}$

S–6–6. Divide: $\dfrac{126}{-21}$

S–6–7. Multiply: $(-3)(-4)(-5)(-6)$

S–6–8. Multiply: $(-29)(-11)(+5)(-13)$

S–6–9. Simplify: $189 - (69 \div 23) \times 2 + 44$

S–6–10. Simplify: $8 \div 8 \times 8 \div 8 \times 8$

S–6–11. Simplify: $-8 - 4(x - 2y) + 6y$

S–6–12. Simplify: $-x + 3y(y - 3) + 7x$

S–6–13. Simplify:
$11 - [6 \div 10(9 - 6 - 3) + 2]$

S–6–14. Simplify:
$35 - [9z - (7z + 2) + 3(5z + 3)]$

S–6–15. Multiply: $y^2 y^3 y^4$

S–6–16. Multiply: $(3x^2)(2x^3)$

S–6–17. Divide: $\dfrac{25x^3y^4}{5x^2y^3}$

S–6–18. Divide: $\dfrac{3xy^5z^2}{18x^2yz^3}$

S–6–19. Divide: $\dfrac{-36a^3b^2c^1}{-6ab^2c^3}$

S–6–20. Multiply: $(-5a^2)^3$

Express the following as algebraic expressions.

S–6–21. The sum of 3, a, and b^3.

S–6–22. The quotient of seven more than h and 7.

S–6–23. The area of a triangle equals one-half the product of the base (b) times the altitude (a).

S–6–24. The side (a) opposite of the 30° angle of 30°–60°–right triangle equals one-half the hypotenuse (c).

S–6–25. The cost to rent money equals the principal (p) times the rate (r) times the time in days (d) divided by 360.

INTRODUCTION

As a child, you learned to crawl, walk, run, and eventually ride a bicycle. Riding a bicycle to get from one place to another was faster and easier than crawling, walking, or running. In a similar way, when you first studied mathematics, you learned whole numbers—to count and then to add, subtract, multiply, and divide. After you developed these basic skills, you moved to more advanced skills. The new skills helped you solve math problems faster and more easily. Algebraic skills will provide solutions to problems that you could not solve with arithmetic. Eventually, you will become skillful working with formulas, which will make your job easier. As you learn the fundamental definitions and operations of algebra, look for ways to use these skills.

DEFINITIONS

Algebra may be considered a language of symbols, numbers, and operations with numbers. An *algebraic expression* uses signs and symbols to represent numbers and quantities. A *numerical algebraic expression* contains only numbers. The quantities separated by a positive or negative sign are called *terms*. Thus $5xy + 3c - z$ is an algebraic expression and the terms are $+5xy$, $+3c$, and $-z$. In the term $6xy$, the 6, x, and y are *factors* of the term. The 6 is the *numerical coefficient*; the x and y are *literal coefficients*. In the example x^5, x is the *base* and 5 is the *exponent* or *power*.

A *monomial* is an algebraic expression that has one term, for example, $5x$. A *binomial*, $4k - j$, has two terms; a *trinomial*, $l + k - j$, has three terms; and a *polynomial*, $k + m + n + s$, has two or more terms. *Like terms* are quantities that have the same literal factors, $4k + 5k - 2k$. *Unlike terms* are terms that have unlike literal factors or powers. Algebra has the same operations as arithmetic (addition, subtraction, multiplication, and division). Multiplication may be expressed in several different ways. The following symbols all indicate multiplication: $a \times b = a \cdot b = (a)(b) = a(b) = (a)b$. The *signs of quantity* indicate whether a number is *positive* (+) or *negative* (−). When someone gives you $5, the number is positive; when you give someone $5, the number is considered negative. A plus sign in front of a number indicates that it is positive; a negative or minus sign in front of a number indicates a negative. When neither a plus nor a minus sign appears in front of a number or quantity, it is considered positive. Often terms are grouped together. Special signs are used to indicate groups. Quantities within a group are to be treated as one quantity. The grouping symbols as follows:

Parentheses	( )
Brackets	[ ]
Braces	{ }
Vinculum or bar	——

Parentheses, brackets, and braces are placed around groups, and a vinculum, or bar, is placed over a group. A vinculum or bar is generally used in writing complex fractions, the division sign, or a radical sign.

ADDITION OF SIGNED VALUES

An algebraic sum may have a positive or negative value. The algebraic sum of two quantities with like signs is the sum of their values, with the same sign in front of that value. The algebraic sum of two or more quantities with unlike signs is the difference of their values, with the sign of the larger quantity. These concepts can be seen in the following examples.

Example 6–1

PROBLEM: Add +3, +2, and +4.

SOLUTION: Plot the given numbers on the number line starting at zero and moving to the right for each number. The number line in Figure 6–1 shows the result.

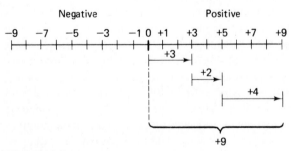

Figure 6–1

Example 6–2

PROBLEM: Find the value of +5, −4, and −3.

SOLUTION: Plot the given numbers on the number line in the order they are given, moving the direction the sign indicates. For example, move to the right from zero for +5 and to the left for −4 and −3. When all the numbers have been plotted, note the distance from zero to the right or left. In this problem the distance to the left of zero is negative 2. Thus the result is −2. See Figure 6–2.

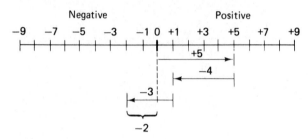

Figure 6–2

MULTIPLICATION AND DIVISION OF SIGNED VALUES

When quantities with unlike signs are multiplied or divided, the results are negative. When terms with like signs are multiplied or divided, the results are positive.

Example 6–3

PROBLEM: If you earned $14 each day for 5 days at a part-time job, how much money did you earn?

SOLUTION: Multiply 5 days × $14.

$$5 \times 14 = 70$$

Thus you have earned $70. See Figure 6–3.

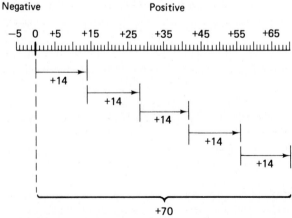

Figure 6–3

Section Two / Basic Algebra

Example 6–4

PROBLEM: If you spend $3 a day for 7 days, what change will there be in your account?

SOLUTION: Because you are spending, you are subtracting from your account; therefore, you multiply -3 times 7.

$$-\$3 \times 7 = -\$21$$

So you would have $-\$21$ in your account. Study the number line in Figure 6–4.

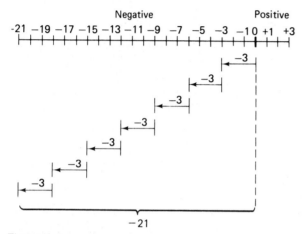

Figure 6–4

Example 6–5

PROBLEM: Multiply -2×-3.

SOLUTION: Subtraction is the reverse of addition; think of this multiplication problem as -3 subtracted from zero 2 times. Remember that like signs result in a positive sign.

$$(-2) \times (-3) = -(-3) - (-3)$$
$$= +3 + 3$$
$$= 6$$

Study the number line shown in Figure 6–5.

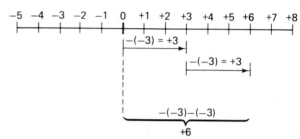

Figure 6–5

Example 6–6

PROBLEM: Divide -42 by 7.

SOLUTION: Write the problem in a fractional form and simplify.

$$\frac{-42}{7} = \frac{-6}{1} = -6$$

EXERCISE 6–1

Solve the following problems according to the signs of operation and indicate the sign of the result.

6–1. $\begin{array}{r} +5 \\ + +3 \\ \hline \end{array}$

6–2. $\begin{array}{r} +8 \\ + -16 \\ \hline \end{array}$

6–3. $\begin{array}{r} -32 \\ + +39 \\ \hline \end{array}$

6–4. $\begin{array}{r} +15 \\ + +3 \\ \hline \end{array}$

6–5. $\begin{array}{r} -9 \\ - -15 \\ \hline \end{array}$

6–6. $\begin{array}{r} -4 \\ + +16 \\ \hline \end{array}$

6-7. $-5 \times +4 =$

6-8. $-6 \times -3 =$

6-9. $+8 \times -3 =$

6-10. $+16 \times +4 =$

6-11. $-4 \times -4 \times -4 =$

6-12. $-5 + 3 - 2 + 4 =$

6-13. $9 - 5 + 6 - 1 =$

6-14. $-11 + 10 - 11 + 10 =$

6-15. $-34 + 72 =$

6-16. $5 \times -3 \times -2 \times -1 =$

6-17. $\dfrac{28}{-7} =$

6-18. $\dfrac{-39}{13} =$

6-19. $\dfrac{-36}{-6} + \dfrac{-6}{+6} =$

6-20. $\dfrac{-20}{+5} + \dfrac{-25}{-5} =$

REMOVAL OF GROUPING SYMBOLS

The removal of grouping symbols is a necessary operation in simplifying an algebraic expression. The sign in front of the grouping determines what operation is to be performed. When there is a negative sign in front of the grouping symbol, change all the signs within the group to the opposite sign and remove the grouping symbol. When a positive sign is in front of the grouping symbol, do not change any of the signs within the grouping symbols as you simplify the expression.

When removing several sets of grouping signs, start with the innermost set of groupings and work to the outermost, making sign changes as necessary. After the grouping symbols have been removed, complete the arithmetic operations. First, perform all multiplication and division operations from left to right. Then perform all addition and subtraction operations, from left to right. Then combine all like quantities.

The *signs of relation* show the value of one term with respect to another. The equal sign (=) means "is equal to." For example, 143 = 143. The equal sign with a slash through it (≠) means "is not equal to." For example, 12 ≠ 11. An arrowhead is used to indicate that a difference exists between two quantities. For example, 5 > 4 indicates that 5 "is greater than" 4 and 6 < 32 indicates that 6 "is less than" 32. The arrowhead always points to the smaller quantity.

Example 6-7

PROBLEM: Evaluate $3(16 \div 2) - 6$.

SOLUTION: First perform the division inside the parentheses.

$$3(16 \div 2) - 6 =$$

$$3\left(\frac{\cancel{16}^{\,8}}{\cancel{2}_{\,1}}\right) - 6 =$$

$$3(8) - 6 =$$

Complete the multiplication, then combine 24 and -6.

$$3(8) - 6 =$$
$$24 - 6 = 18$$

Example 6-8

PROBLEM: Simplify $5 + \{3x - [2 - (5 + 5x)]\}$.

SOLUTION: First, remove the innermost parentheses by changing every sign within the parentheses, because it is preceded by a negative sign.

$$5 + \{3x - [2 - (5 + 5x)]\} =$$
$$5 + \{3x - [2 - 5 - 5x]\} =$$

Then remove the brackets by changing every sign within the brackets as indicated.

$$5 + \{3x - [2 - 5 - 5x]\} =$$
$$5 + \{3x - 2 + 5 + 5x\} =$$

Remove the braces but do not change any signs because the braces are preceded by a positive sign.

$$5 + \{3x - 2 + 5 + 5x\} =$$
$$5 + 3x - 2 + 5 + 5x =$$

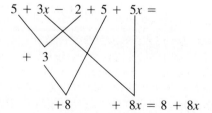

Example 6-9

PROBLEM: Simplify

$$\{24 \div 6 \div 2 - (3x - 6[3 - 2x]) + 4\}$$

SOLUTION: First, remove the innermost brackets by multiplying the enclosed terms by -6.

$$\{24 \div 6 \div 2 - (3x - 6[3 - 2x]) + 4\} =$$
$$\{24 \div 6 \div 2 - (3x - 18 + 12x) + 4\} =$$

Second, remove the parentheses by changing every sign enclosed by the parentheses because of the negative sign in front of the parentheses.

$$\{24 \div 6 \div 2 - (3x - 18 + 12x) + 4\} =$$
$$\{24 \div 6 \div 2 - 3x + 18 - 12x + 4\} =$$

Then perform the division operations, moving from left to right as shown.

$$\{24 \div 6 \div 2 - 3x + 18 - 12x + 4\} =$$
$$\left\{\frac{24}{6} \div 2 - 3x + 18 - 12x + 4\right\} =$$
$$\{4 \div 2 - 3x + 18 - 12x + 4\} =$$
$$\left\{\frac{4}{2} - 3x + 18 - 12x + 4\right\} =$$
$$\{2 - 3x + 18 - 12x + 4\} =$$

Remove the braces but do not change any signs. Since there is no sign in front of the braces, it is understood to be positive.

$$2 - 3x + 18 - 12x + 4 =$$

Then combine like quantities.

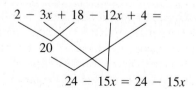

Chapter 6 / Definitions and Basic Operations of Algebra

EXERCISE 6-2

Simplify the following by performing the operations indicated. Show all necessary work.

6-21. $3 \times 4 + 5 \times 6 =$

6-22. $8 \times 7 - 6 \times 6 =$

6-23. $(16 \div 2) - 8 =$

6-24. $60 \div (5 \times 3) =$

6-25. $60 \div 5 \times 3 =$

6-26. $8 \div 8 \times 8 \div 8 =$

6-27. $(8 \div 8) \times (8 \div 8) =$

6-28. $24 - (90 \div 18) + 5 =$

6-29. $390 \div 13 - 45 + 10 =$

6-30. $x + y - z + z - x + y =$

6-31. $a + b - (3a + 5b) - \{a - b\} =$

6-32. $(6a + 2b) - (4b - 3b - 2b) =$

6–33. $3k + 2 - [2j + 5 - 6j + 3k - 3] =$

6–34. $4r - 8y - (3r - 4y) + 4r - 32y =$

6–35. $\{(r + 3t - 8) - (r - 3t + 8)\} =$

6–36. $\{-3a - 2b - (c - a)\} =$

6–37. $128 \div 2 \div 2 \div 2 \div 2 =$

6–38. $4a - [-5b - (-12b + 3a)] - \{a + 3b\} =$

6–39. $7b - 4\{2b + 6k - [-2b + 6k] - 4b\} =$

6–40. $8x - [4y - (-3x + 3y) - 2x] =$

WORKING WITH EXPONENTS

An *exponent* is a number written above and to the right of a quantity and indicates how many times the *base number* is used as a factor. Exponents are used to abbreviate algebraic expressions: for example, $k \cdot k \cdot k \cdot k = k^4$. The letter k is the base number or base quantity, and 4 is the exponent. Study the following examples, which show the laws of exponents.

1. When multiplying quantities with like bases containing exponents, add the exponents.

$$(j^3)(j^2) = j^{3+2} = j^5$$

or

$$(j \cdot j \cdot j)(j \cdot j) = j \cdot j \cdot j \cdot j \cdot j = j^5$$

Thus

$$(a^m)(a^n) = a^{m+n}$$

or

$$(3)^3(3)^2 = (3)^5 = (3)(3)(3)(3)(3) = 243$$

2. When dividing quantities with like bases containing exponents, subtract the exponent of the denominator from the exponent of the numerator.

$$\frac{y^5}{y^2} = \frac{\overset{1}{\cancel{y}} \cdot \overset{1}{\cancel{y}} \cdot y \cdot y \cdot y}{\underset{1}{\cancel{y}} \cdot \underset{1}{\cancel{y}}}$$

or

$$\frac{y^5}{y^2} = y^{5-2} = y^3$$

Thus

$$\frac{a^m}{a^n} = a^{m-n}$$

or

Chapter 6 / Definitions and Basic Operations of Algebra

$$\frac{(3)^5}{(3)^2} = (3)^{5-2}(3)^3 = 27$$

3. A quantity with an exponent of zero is equal to 1.

$$1 = \frac{1}{1} = \frac{\overset{1}{\cancel{a}}\cdot\overset{1}{\cancel{a}}\cdot\overset{1}{\cancel{a}}\cdot\overset{1}{\cancel{a}}}{\underset{1}{\cancel{a}}\cdot\underset{1}{\cancel{a}}\cdot\underset{1}{\cancel{a}}\cdot\underset{1}{\cancel{a}}} = \frac{a^4}{a^4} = a^{4-4} = a^0 = 1$$

Thus

$$\frac{a^m}{a^m} = a^{m-m} = a^0$$

The following examples illustrate how exponents should be solved.

Example 6–10

PROBLEM: Multiply $x^3 \cdot x^6$.

SOLUTION: $(x^3)(x^6) = x^{3+6} = x^9$ because

$$\underbrace{x \cdot x \cdot x}_{(x^3)} \cdot \underbrace{x \cdot x \cdot x \cdot x \cdot x \cdot x}_{(x^6)} = x^9$$

When you multiply quantities with the same bases, you add the exponents of like terms.

EXERCISE 6–3

Perform the operations indicated and evaluate when necessary.

6–41. $m^3 \cdot m^4 =$

6–43. $5^4 \cdot 5^1 =$

6–45. $a^5 \cdot a^{10} \cdot a^{15} =$

Example 6–11

PROBLEM: Divide: $y^8 \div y^5$.

SOLUTION: $y^8 \div y^5 = y^{8-5} = y^3$ because

$$\frac{y^8}{y^5} = \frac{\overset{1}{\cancel{y}}\cdot\overset{1}{\cancel{y}}\cdot\overset{1}{\cancel{y}}\cdot\overset{1}{\cancel{y}}\cdot\overset{1}{\cancel{y}}\cdot y \cdot y \cdot y}{\underset{1}{\cancel{y}}\;\underset{1}{\cancel{y}}\;\underset{1}{\cancel{y}}\;\underset{1}{\cancel{y}}\;\underset{1}{\cancel{y}}} = y^3$$

When you divide the bases, subtract the exponents (denominator from numerator) of the like terms.

Example 6–12

PROBLEM: Simplify: $(6x^2y^3)^2$.

SOLUTION:

$$(6x^2y^3)^2 = (6x^2y^3)(6x^2y^3)$$
$$= (6)(6)x^{2+2}y^{3+3}$$
$$= 36x^4y^6$$

When you raise to a power, quantities, multiply the numerical factors and add the exponents of the like bases.

6–42. $x^3 \cdot x^4 \cdot x^2 =$

6–44. $3 \cdot 3 \cdot 3^2 =$

6–46. $xy \cdot xy \cdot xy =$

6–47. $x^3y^3 \cdot x^4y^4 \cdot x^1y^2 =$

6–48. $(xyz)(xyz) =$

6–49. $(xy)^3 =$

6–50. $(-4)^2(-4)^1(-4)^3 =$

6–51. $\dfrac{x^5}{x^4} =$

6–52. $\dfrac{x^3y^2}{x^2y^2} =$

6–53. $(x^3)^4 =$

6–54. $(x^3y^2)^3 =$

6–55. $-(x^1y^2z^3)^4 =$

6–56. $\left(\dfrac{y^3}{y^2}\right)^3 =$

6–57. $\left(\dfrac{2x^2}{4y^3}\right)^2 =$

6–58. $\left(\dfrac{3^2x^2y^3}{x^1y^2}\right)^2 =$

6–59. $\dfrac{12r^1s^2t^3}{6rst} =$

6–60. $\left(\dfrac{16x^2r^4s^5}{2^4x^2s^5r^4}\right)^2 =$

Chapter 6 / Definitions and Basic Operations of Algebra

CONVERTING WRITTEN PROBLEMS INTO EQUATIONS

One of the more difficult tasks to perform in mathematics is to change a written problem to a formula or an equation. Some of the reasons are (1) a lack of understanding of the problem and what is to be answered; (2) a lack of the skill to translate words and ideas into symbols; or (3) a lack of everyday application and use of the formulas or equations. Mathematical operations are often expressed by words and phrases that mean the same thing. Some of these are as follows:

Addition: sum; combine; total; add.
Subtraction: find the difference; subtracted from; is how much greater; decreased by; the difference between; is less than.
Multiplication: find the product; square the quantity; cube the quantity; double the quantities.
Division: divided by; find the quotient.

As you work the following word problems, keep in mind what the words tell you to do, how the words are translated into symbols, and how the problem is solved.

Example 6–13

PROBLEM: Translate to an equation: The sum of three numbers doubled is 38.

SOLUTION: Define the numbers:

$$x = \text{the first number}$$
$$y = \text{the second number}$$
$$z = \text{the third number}$$

Sum means to add and *double* means to multiply all the numbers by 2. Therefore,

$$2(x + y + z) = 38$$

Sum means to add and *double* means to multiply all the numbers by 2. Therefore,

$$2(x + y + z) = 38$$

Example 6–14

PROBLEM: Translate to an equation: The distance traveled depends on the time and the speed of travel.

SOLUTION: Define the quantities involved:

$$d = \text{the distance}$$
$$t = \text{the time traveled}$$
$$r = \text{the rate of travel}$$

Since the distance traveled depends on the hours traveled times the rate of travel, we have

$$d = rt$$

Example 6–15

PROBLEM: Translate to an equation: The volume of a cone is $\frac{1}{3}$ of the radius times the radius times the height times π.

SOLUTION: Define the given quantities: V = volume, r = radius, h = height, π = 3.142. Thus

$$V = \frac{1}{3}(r)(r)(h)(3.142)$$

or

$$V = \frac{3.142 r^2 h}{3}$$

EXERCISE 6–4

Translate the following word problems into formulas or equations. Review the previous examples if necessary.

6–61. Find the sum of two numbers minus a third number.

6–62. Subtract y from the sum of x^3 and z^4.

6–63. The product of two numbers divided by a third number equals 6.

6–64. To convert liters to quarts, multiply by 1.05.

6–65. To convert feet to inches, multiply by 12.

6–66. The perimeter of a square equals the sum of the lengths of the sides.

6–67. The area of a square equals the length of one side times the length of another side.

6–68. The perimeter of a rectangle equals the sum of the lengths of the sides.

6–69. The circumference of a circle equals the diameter times π.

6–70. Miles per gallon equals the miles driven divided by the number of gallons used.

6–71. Four times k plus 2 times p equals f.

6–72. The surface area of a cube equals 6 times the area of each face.

6–73. The sum of three consecutive numbers is 37.

6–74. Square the sum of a plus b minus c cubed.

6–75. The quotient of *j* and *k* times *l* equals *t* minus *s*.

6–76. The number of gallons of water in a container is equal to the product of length, width, height, and 7.5.

6–77. The hypotenuse squared of a right triangle equals the sum of the lengths of the other two sides squared.

6–78. The strength of a board depends on the product of the width and the depth squared, divided by the length.

6–79. The weight in pounds of water in a tank equals the product of the length, width, height, and 62.5.

6–80. The cost of carpeting a room depends on the area in square yards times the cost per square yard. Determine the cost if the room's dimensions are *x* yards by *y* yards.

THINK TIME

One of the symbols frequently used in mathematics is zero, 0. Several facts about zero are important when doing algebraic operations.
1. Zero is the point between positive and negative numbers. It is the only quantity that is neither positive nor negative.
2. When determining the powers of 10, each additional zero to the right of the number 10 indicates an additional power of 10. For example:

   $10 = 10^1, \quad 100 = 10^2, \quad 1000 = 10^3$

3. The addition or subtraction of zero to a number results in the given number.
4. When any number is multiplied by 0, the product is 0.
5. When 0 is divided by any other number, the quotient is 0.
6. Division of any number by 0 is *impossible* or *undefined*.
7. A quantity to the zero power equals 1.

PROCEDURES TO REMEMBER

1. The algebraic sum of quantities with *like* signs is the sum of their values, with the same sign in front of the combined quantities.
2. The algebraic sum of quantities with *unlike* signs is the difference of their values, with the sign of the greater value in front of the combined quantities.
3. When terms with like signs are multiplied or divided, the results are positive.
4. When terms with unlike signs are multiplied or divided, the results are negative.
5. When removing grouping symbols with a positive sign in front, do not change any signs within the group.
6. When removing grouping symbols with a negative sign in front, change each of the signs within the group to the opposite sign indicated.
7. When removing several sets of grouping symbols, start with the innermost set of groupings and work to the outermost, making sign changes as necessary.

8. After grouping signs have been removed and all necessary sign changes have been made, first perform all multiplication and division operations from left to right and then perform all addition and subtraction operations, moving from left to right.
9. When multiplying quantities with like bases containing exponents, add the exponents.
10. When dividing quantities with like bases with exponents, subtract the exponent of the denominator from the exponent of the numerator. When a quantity is moved from the numerator to the denominator, and vice versa, change the sign of the exponent.

CHAPTER SUMMARY

1. An *algebraic expression* is an expression that uses signs and symbols to represent numbers and quantities.
2. A *numerical algebraic expression* is made up entirely of numbers.
3. A *term* is part of an expression and is separated from other terms by a positive or negative sign.
4. One or more *factors* may be combined to form a term.
5. A *numerical coefficient* refers to the numerical part of a term.
6. A *literal coefficient* refers to literal parts of a term.
7. The *base* is the quantity that is used as a factor when working with exponents.
8. An *exponent* or *power* indicates the number of times the base is multiplied by itself.
9. A *monomial* is an algebraic expression that has one term.
10. A *binomial* is an algebraic expression that has two terms.
11. A *trinomial* is an algebraic expression that has three terms.
12. A *polynomial* is an algebraic expression that has two or more terms.
13. *Like* terms are quantities that have the same literal factors.
14. *Unlike* terms are quantities that have unlike literal factors.
15. *Signs of quantity* indicate whether a number is positive or negative.
16. *Parentheses, brackets, braces,* and *vincula* or *bars* are used to group terms.
17. *Signs of operation* indicate what operation should be performed.
18. The equal sign ($=$) means "is equal to."
19. The equal sign with a slash through it ($\neq$) means "is not equal to."
20. An arrowhead is used to indicate that a difference exists between two quantities. For example, $8 > 4$ indicates that 8 "is greater than" 4, and $8 < 9$ indicates that 8 "is less than" 9. The arrowhead always points to the smaller quantity.
21. A quantity with a zero exponent is equal to 1.
22. Addition means to "combine," "total," "add," and "sum."
23. Subtraction means to "find the difference," "subtracted from," "is how much greater than," "decreased by," and "the difference between."
24. Multiplication means to "find the product," "square the quantity," "cube the quantity," and "double the quantities."
25. Division means to "divide by" and "find the quotient."

CHAPTER TEST

T–6–1. Add: -35
$\phantom{\text{Add: }}-19$

T–6–2. Add: -140
$\phantom{\text{Add: }}93$
$\phantom{\text{Add: }}120$
$\phantom{\text{Add: }}-62$

T–6–3. Subtract: -23
$\phantom{\text{Subtract: }}+18$

T–6–4. Subtract: -98
$\phantom{\text{Subtract: }}-46$

T–6–5. Multiply: $(-4)(+5)(-6) =$

T–6–6. Multiply: $(-30)(-2)(-2) =$

T–6–7. Divide: $\dfrac{-111}{+3}$

T–6–8. Divide: $\dfrac{+27}{-3}$

T–6–9. Subtract: $\begin{array}{r} -66ab \\ +76ab \end{array}$

T–6–10. Simplify:
$84 + 36 \times 8 \div 4 \times 2 - 12 =$

T–6–11. Simplify:
$8 - 0(6 \div 3) - 16 \div 2 =$

T–6–12. Simplify:
$3x - 7 - 2(x - 5) + 15 =$

T–6–13. Simplify:
$4 + \{10 - [-6 - (-7 + 4) - 2]\} =$

T–6–14. Simplify:
$(5x - 2) - 2\{3x - [5x + 7(3x - 4) + 3]\} =$

T–6–15. Multiply: $-b^3 \cdot b^5 =$

T–6–16. Multiply: $(12x^3)(12x^2y^3) =$

T-6-17. Divide: $\dfrac{-64x^3y^5z^2}{-16x^3y^4z} =$

T-6-18. Divide: $\dfrac{45j^3k^3l^3}{-5j^3k^2l^1} =$

T-6-19. Solve: $(-4j^3)^4 =$

Express the following as algebraic expressions.

T-6-20. The difference between x and y divided by r^3.

T-6-21. The sum of a and b minus c multiplied by x^2 divided by y^4 minus 7.

T-6-22. The product of 36 minus z times j.

T-6-23. The area of a trapezoid equals one-half the sum of the bases (b_1 and b_2) times the altitude (a).

T-6-24. The perimeter of a 60°–60°–60° triangle equals three times one side (s).

T-6-25. The cost (C) to operate a car equals the cost per gallon (c) times the gallons (g) plus insurance (i) plus oil (o) plus tires (t) plus maintenance $c(m)$ plus other expenses.

7

Simple Equations and Formulas

OBJECTIVES

1. To solve simple equations using the algebraic axioms or the procedures of addition, subtraction, multiplication, and division.
2. To solve problems using formulas.

SELF-TEST

This basic equations test will determine what you need to study in this chapter. Solve for the unknown, showing all necessary work. Check your work.

S–7–1. $x - 5 = 3$

S–7–2. $4x + 5 = 17$

S–7–3. $x + 3 = 10$

S–7–4. $2x + 3 = x + 4$

S–7–5. $4b = 36$

S–7–6. $17x - 7x = x + 18$

S–7–7. $4x + 5 - 7 = 2x + 6$

S–7–8. $9y - 19 = y - 2y + 11$

S–7–9. $4x - 10 = 2x + 2$

S–7–10. $9x + 9 = -3x + 15$

S–7–11. $300x - 250 = 50x + 750$

S–7–12. $2.5x + 0.5x = 1.5x + 1.5$

S–7–13. $x + 2x + 3 - 4x = 5x - 9$

S–7–14. $2y + 3y - 4 = 5y + 6y - 16$

S–7–15. $75z - 150 = 80z - 300$

S–7–16. $(4x + 6) - 2x = (x - 6) + 24$

S–7–17. $15y - [3 - (4y + 4) - 57] = -(2 + y)$

S–7–18. $4t - (12t - 24) + 38t - 39 = 0$

S–7–19. $47r - 17 = 235 - 37r$

S–7–20. $ax + b = 7$. Solve for x.

S–7–21. $3(b - y) = 5(b - 2y)$. Solve for y.

S–7–22. $5(x - 2b) = -3x + 2(x + 2b)$. Solve for x.

Chapter 7 / Simple Equations and Formulas

S–7–23. $\frac{3}{4}(3a - 2x) = \frac{1}{4}x + \frac{1}{4}a$. Solve for x.

S–7–24. $5.2(x + 3) + 3.7(2 - x) = 3$. Solve for x.

S–7–25. $(x + 3)^2 = (x - 4)^2$. Solve for x.

INTRODUCTION

In the modern world we use many formulas to solve problems. As technology advances, many more formulas will be developed to solve new problems. It is essential that technical workers have knowledge of formulas and methods used to solve problems. In this chapter we give explanations, illustrations, and an opportunity to work with formulas to become skilled in their use. Many people fail to understand the relationship between formulas and equations. What a technical worker calls a formula, the mathematician calls a literal equation.

As you learn the fundamental operations of equations, be aware that these are the same procedures used to solve formulas used in business and industry.

DEFINITIONS

An *equation* is a statement with two algebraic expressions that are equal. For example, $3x = 12$ has two algebraic expressions, $3x$ and 12, that are equal. A *formula* is a rule or procedure used in mathematics, science, or industry to solve problems. Formulas are generally composed of letters, symbols, and constant terms. *Literal quantities* are letters that represent numbers or unknown quantities.

Historically, the symbol for justice and equality has been the balance scale, often called the "scales of justice." The scales of justice can serve as an aid to understanding equations. See Figure 7–1. The balance point

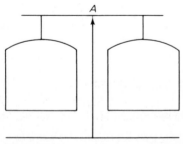

Figure 7–1

A is equivalent to the equal sign, $=$, of an equation. To keep the scales balanced, if you add, subtract, multiply, or divide on one side of the scale, you must do the same operation on the other side of the scale. Failure to do the same operation on each side will result in an imbalance or error.

Study Figure 7–2. Note in each equation, what is

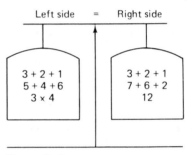

Figure 7–2

given on the left side is equal to what is given on the right side. The quantities on the left side of the equation are known as the *left member*. The quantities on the right side are known as the *right member*. The equal sign in the center is the *sign of equality* or the balance point of the equation. Remember, whatever mathematical operation you do on one side of the equation must be done on the other side. Failure to practice this principle will result in an "injustice." On the job, incorrect use of a formula may result in an injustice to yourself or to your customer.

The procedures used in solving equations are the same as those used in solving formulas. Using the skills you will learn solving equations will help you solving formulas.

The four basic operations used in solving equations are the same as the fundamental operations of arithmetic: addition, subtraction, multiplication, and division. However, when used to solve equations, they are called the *axioms* or *rules* of algebra.

Study Figures 7–3 to 7–6. Note that the balance scale remains balanced as weights are added, subtracted, multiplied, or divided in equal amounts.

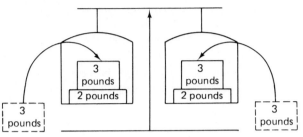

Figure 7–3   Add equal weight to each side.

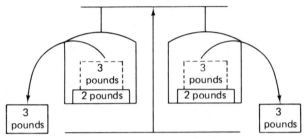

Figure 7–4   Subtract equal weight from each side.

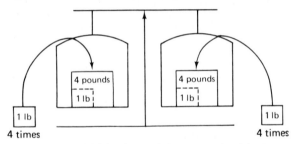

Figure 7–5   Multiply the weight on each side by an equal quantity.

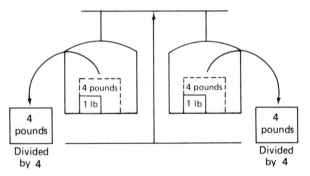

Figure 7–6   Divide the weight on each side by an equal quantity.

The solution or value of the unknown quantity is called the *root* of the equation. When the root is found, the problem has been solved or the *solution set* has been determined. Always check your solutions to prevent costly mistakes or errors. Always check your solutions to be certain that you have the correct balance or that both sides of the equation are equal.

The following examples illustrate the basic axioms of algebra: addition, subtraction, multiplication, and division. Note that there is often more than one way to solve an equation.

SOLVING EQUATIONS USING ADDITION OR SUBTRACTION

Addition Axiom

The quantity added to the left side of the equation must also be added to the right side.

Example 7–1

PROBLEM: Find the value of X in the equation $X - 3 = 2$.

SOLUTION A: Solve this equation by *adding* 3 to each side of the equation.

$$\begin{array}{rl} X - 3 = & 2 \\ + 3 = & +3 \\ \hline X = & 5 \end{array}$$

CHECK: Substitute 5 for X in the original equation and evaluate.

$$\begin{array}{r} X - 3 = 2 \\ 5 - 3 = 2 \\ 2 = 2 \end{array}$$

SOLUTION B: This problem could be solved by transposition, which is the process of moving a quantity from one side of the equal sign to the other side. When a quantity is moved to the opposite side, its sign is changed. This procedure is equivalent to adding the same quantity to each side of the equation. Thus you could solve this equation by moving the -3 to the other side of the equal sign and changing it to $+3$. Thus

$$\begin{array}{r} X - 3 = 2 \\ X = 2 + 3 \\ X = 5 \end{array}$$

Subtraction Axiom

The quantity that is subtracted from the left side of the equation must also be subtracted from the right side.

Chapter 7 / Simple Equations and Formulas

Example 7-2

PROBLEM: Find the value of X in the equation $X + 4 = 6$.

SOLUTION A: Solve this equation by *subtracting* 4 from each side of the equation.

$$\begin{aligned} X + 4 &= 6 \\ -4 &= -4 \\ \hline X &= 2 \end{aligned}$$

CHECK: Substitute 2 for X in the original equation and evaluate.

$$X + 4 = 6$$
$$2 + 4 = 6$$

SOLUTION B: This problem may be solved by transposition, by moving the $+4$ to the opposite side of the equation sign and changing its sign from plus to minus.

Thus

$$X + 4 = 6$$
$$X = 6 - 4$$
$$X = 2$$

Example 7-3

PROBLEM: Find the value of X in the equation $8X - 5 = 7X + 19$.

SOLUTION A: Solve this equation by *subtracting* $7X$ from each side of the equation and then *adding* 5 to each side of the equation.

$$\begin{aligned} 8X - 5 &= 7X + 19 \\ -7X &= -7X \\ \hline X - 5 &= +19 \end{aligned}$$

$$\begin{aligned} X - 5 &= 19 \\ +5 &= +5 \\ \hline X &= 24 \end{aligned}$$

CHECK: Substitute 24 for X in the original equation and evaluate.

$$8X - 5 = 7X + 19$$
$$8(24) - 5 = 7(24) + 19$$
$$192 - 5 = 168 + 19$$
$$187 = 187$$

SOLUTION B: This problem may be solved by transposition: $+7X$ is moved to the opposite side of the equal sign and its sign changed from positive to negative. Then the -5 is moved to the opposite side of the equal sign and its sign changed from negative to positive. Observe:

$$8X - 5 = 7X + 19$$
$$8X - 7X - 5 = +19$$
$$8X - 7X = 19 + 5$$
$$X = 24$$

Example 7-4

PROBLEM: Find the value of X in the equation $3(3X - 4) = 4(2X + 6) - 20$.

SOLUTION: Remove the parentheses by completing the multiplication indicated.

$$3(3X - 4) = 4(2X + 6) - 20$$
$$9X - 12 = 8X + 24 - 20$$

Then move -12 and $+8X$ to the opposite side of the equation and combine the quantities. Remember when moving a quantity from one side of the equal sign to the other to change the sign on the quantity.

$$9X - 12 = 8X + 24 - 20$$
$$9X - 8X = 12 + 24 - 20$$
$$X = 16$$

CHECK: Substitute 16 for X in the original equation and evaluate.

$$3(3X - 4) = 4(2X + 6) - 20$$
$$3[3(16) - 4] = 4[2(16) + 6] - 20$$
$$3[48 - 4] = 4[32 + 6] - 20$$
$$3[44] = 4[38] - 20$$
$$132 = 152 - 20$$
$$132 = 132$$

EXERCISE 7-1

Solve the following equations using either method. Check your solutions by substituting your answer for the unknown in the original equation. Study the examples if you need help. Remember that the same operation must be done on both sides of the equation.

7-1. $X + 3 = 5$

7-2. $X - 6 = 10$

7-3. $3X - 4X = -2X + 10$

7-4. $8y - 12 = 7y - 11$

7-5. $X - 7 = 13$

7-6. $3 + a = 10$

7-7. $k - 12 = 3$

7-8. $r - 4 = -18$

7-9. $X + 14 = 8$

7-10. $2(X + 10) = X + 30$

7-11. $3(X - 1) + 2X = 2(3X + 4)$

7-12. $4(3 - 2X) = 7(5 - X)$

Chapter 7 / Simple Equations and Formulas

7–13. $3(3X + 3) = 4(2X - 4)$

7–14. $3(X - 8) = 2(X - 14)$

7–15. $6(2x) - 16 = 4x + 32$

7–16. $2(18 - 4x) = 4(2x + 4)$

7–17. $5(2x - 5) = 4x + 5$

7–18. $8(x - 4) = 2x + 10$

7–19. $4(x + 8) = 6(x + 6) + 4$

7–20. $7(x - 1) - 2 = 5(x + 1)$

SOLVING EQUATIONS USING DIVISION OR MULTIPLICATION

Not all equations can be solved by addition and subtraction procedures. For example, in the equation $5X = 15$, the unknown is greater than 1 unit; thus the unknown should be divided by 5 so that it becomes 1 unit. However, remember to divide both sides of the equation by 5.

Division Axiom

Each side of the equation must be divided by the same quantity.

Example 7–5

PROBLEM: Find the value of X in the equation $5X = 15$.

SOLUTION: Divide both sides of the equation by 5.

$$5X = 15$$
$$\frac{\cancel{5}X}{\cancel{5}} = \frac{\cancel{15}^3}{\cancel{5}}$$
$$X = 3$$

CHECK: Substitute $X = 3$ into the original equation and evaluate.

$$5X = 15$$
$$5(3) = 15$$
$$15 = 15$$

To solve $\frac{2}{3}X = 12$, an equation where the unknown is a fraction, multiply both sides by $\frac{3}{2}$, the reciprocal of $\frac{2}{3}$. If the unknown is a fraction, multiply *both* sides of the equation by the reciprocal of the fraction. Check by substituting the root into the original equation.

> **Multiplication Axiom**
> Each side of the equation must be multiplied by the same quantity.

Example 7–6

PROBLEM: Find the value of X is the equation $\frac{2}{3}X = 12$.

SOLUTION: Multiply both sides of the equation by $\frac{3}{2}$, the reciprocal of $\frac{2}{3}$.

$$\frac{2}{3}X = 2$$

$$\left(\frac{3}{2}\right)\frac{2}{3}X = \frac{12}{1}\left(\frac{3}{2}\right)$$

$$X = 18$$

CHECK: Substitute $X = 18$ into the original equation and evaluate.

$$\frac{2}{3}X = 12$$

$$\frac{2}{3}\left(\frac{18}{1}\right) = 12$$

$$12 = 12$$

SOLVING EQUATIONS USING COMBINED METHODS

Study the following examples, which use more than one of the algebra axioms.

Example 7–7

PROBLEM: Find the value of X in the equation

$$8X - 5(4X + 3) = (-3) - 4(2X - 7)$$

SOLUTION: Remove the parentheses.

$$8X - 20X - 15 = (-3) - 8X + 28$$

Combine the like quantities as indicated.

$$8X - 20X - 15 = (-3) - 8X + 28$$
$$-12X - 15 = -8X + 25$$

Transpose the quantities as indicated by the grouping symbols.

$$-12X - 15 = -8X + 25$$
$$-25 - 15 = -8X + 12X$$

Combine like quantities as indicated by the grouping symbols.

$$-25 - 15 = -8X + 12X$$
$$-40 = 4X$$

Then divide both sides by 4.

$$-40 = 4X$$

$$\frac{-40}{4} = \frac{4X}{4}$$

$$-10 = X$$

CHECK: Substitute $X = -10$ into the original equation and evaluate.

$$8X - 5(4X + 3) = (-3) - 4(2X - 7)$$
$$8(-10) - 5[4(-10) + 3] = (-3) - 4[2(-10) - 7]$$
$$-80 - 5[-40 + 3] = (-3) - 4[-20 - 7]$$
$$-80 - 5[-37] = (-3) - 4[-27]$$
$$-80 + 185 = -3 + 108$$
$$105 = 105$$

Example 7–8

PROBLEM: Find the value of k in the equation

$$4k = 3[2k - 4(k - 2)] = 72 - 6k$$

SOLUTION: Remove the inside parentheses.

$$4k = 3[+2k - 4x + 8] = 72 - 6k$$

Combine like quantities.

$$4k = 3[-2k + 8] = 72 - 6k$$

Then remove the outside parentheses.

$$4k - 6k + 24 = 72 - 6k$$

Combine like quantities.

$$-2k + 24 = 72 - 6k$$

Then transpose quantities as necessary so that the positive unknown is on the left side of the equation.

$$-2k + 6k = 72 - 24$$

Combine like quantities.

$$4k = 48$$

Divide both sides by $+4$.

$$\frac{\cancel{4}k}{\cancel{4}} = \frac{\cancel{48}^{12}}{\cancel{4}}$$

$$k = 12$$

CHECK: Substitute $k = 12$ in the original equation and evaluate.

$$4k + 3[2k - 4(k - 2)] = 72 - 6k$$
$$4(12) + 3[2(12) - 4(12 - 2)] = 72 - 6(12)$$
$$48 + 3[24 - 4(10)] = 72 - 72$$
$$48 + 3[24 - 40] = 0$$
$$48 + 3[-16] = 0$$
$$48 - 48 = 0$$
$$0 = 0$$

Example 7-9

PROBLEM: Find the value of the y in the equation

$$16.5 - 1.5(2y - 0.5) - 15.6 + 2.1(y + 0.3) = 0.03$$

SOLUTION: Remove the parentheses.

$$16.5 - 3.0y + 0.75 - 15.6 + 2.1y + 0.63 = 0.03$$

Combine like quantities.

$$-0.9y + 2.28 = 0.003$$

Move 2.28 to the right side of the equation and remember to change the sign.

$$-0.9y = 0.03 - 2.28$$

Combine like quantities.

$$-0.9y = -2.25$$

Divide both sides by -0.9.

$$\frac{\cancel{-0.9}y}{\cancel{-0.9}} = \frac{\cancel{-2.25}^{2.5}}{\cancel{-0.9}}$$

$$y = 2.5$$

CHECK: Substitute $y = 2.5$ into the original equation and evaluate.

$$16.5 - 1.5(2y - 0.5) - 15.6 + 2.1(y + 0.3) = 0.03$$
$$16.5 - 1.5(2(2.5) - 0.5) - 15.6 + 2.1(2.5 + 0.3) = 0.03$$
$$16.5 - 1.5(5 - 0.5) - 15.6 + 2.1(2.8) = 0.03$$
$$16.5 - 1.5(4.5) - 15.6 + 5.88 = 0.03$$
$$16.5 - 6.75 - 15.6 + 5.88 = 0.03$$
$$0.03 = 0.03$$

Example 7-10

PROBLEM: Find the value x in the equation

$$(x + 3)(x + 3) - 8 = (7 + x)(4 - x) + 2x^2$$

SOLUTION: Remove the parentheses by multiplying as shown and combine like quantities.

$$(x + 3)(x + 3) - 8 = (7 + x)(4 - x) + 2x^2$$
$$x^2 + 3x + 3x + 9 - 8 = 28 - 7x + 4x - x^2 + 2x^2$$
$$x^2 + 6x + 1 = 28 - 3x + x^2$$

Move the unknowns to the right side of the equal sign and combine the quantities.

$$x^2 + 6x + 1 = 28 - 3x + x^2$$
$$x^2 - x^2 + 6x + 3x = 28 - 1$$
$$0x^2 + 9x = 27$$

Solve for x by dividing both sides by 9.

$$\frac{\cancel{9}x}{\cancel{9}} = \frac{\cancel{27}^{3}}{\cancel{9}}$$

Thus $x = 3$.

CHECK: Substitute $x = 3$ into the original equation and evaluate.

$$(x + 3)(x + 3) - 8 = (7 + x)(4 - x) + 2x^2$$
$$(3 + 3)(3 + 3) - 8 = (7 + 3)(4 - 3) + 2(3)^2$$
$$(6)(6) - 8 = (10)(1) + 2(9)$$
$$36 - 8 = 10 + 18$$
$$28 = 28$$

EXERCISE 7–2

Some of the steps in previous problems may be combined when you become more comfortable at solving equations. However, while learning, it is a good idea to show each operation you perform.

7–21. $3a - 9 = 9$

7–22. $5L + 3 = 2 + 4L$

7–23. $\frac{2}{3}X - 9 = \frac{1}{3}X$

7–24. $k - 10 = 5 + 4k$

7–25. $2t + 8 = t + 13$

7–26. $4d + 15 = 6d + 5$

7–27. $12R - 18 + 42 = 10(R + 1)$

7–28. $4X + 8 = 2(4X + 2)$

7–29. $11X + 3 - 4X = 16 - 2X + 2$

7–30. $18 - 5(3 - 2z) = z - 4(z + 9)$

7–31. $4 + 3(J - 7) = 16 + 2(J + 1) + 3J$

7–32. $4(a - 5) - 3(a - 2) = 2(a - 1)$

Chapter 7 / Simple Equations and Formulas

7-33. $(2x)^2 + 2(x) + 1 = 4x^2 + 9$

7-34. $5X - (3X - 2) = 10$

7-35. $(a + 1)(a - 2) = a^2 + 5$

7-36. $\frac{1}{5}n + 7 = 13$

7-37. $\frac{1}{3}X + \frac{1}{4}X = \frac{7}{2}$

7-38. $10(y - 2) - 10(2 - y) = 4y - 40$

7-39. $(b + 4)(5 - 2b) - b(10 - 2b) = -6$

7-40. $(a + 5)(a - 4) + 4a^2 = (5a + 3)(a - 4) + 2(a - 4) + 64$

7-41. $\frac{3}{4}x + \frac{1}{3}x = 4x - 35$

7-42. $9(k + 2) - (2k + 19) = 4k - 10$

7-43. $(h + 6)(h - 6) = (9 - h)(4 + 3h) + 4h^2 - 3$

7-44. $(2c + 3)(c - 2) - c^2 = (c + 8)(c + 2)$

7-45. $(y + 1)(y^2 - y + 1) = y^3 - 8y - 31$
(*Hint:* Perform the multiplication as shown in Example 7–10.)

Be sure that you checked all your solutions. If not—
CHECK THEM!

LITERAL EQUATIONS

A *literal equation* is an equation with some or all of the quantities represented by letters. Letters from the beginning of the alphabet usually represent known quantities, and letters toward the end of the alphabet represent unknown quantities. When solving an equation for x, y, or z and the rest of the equation contains literal factors, you are solving a literal equation.

The same procedures are used to solve literal equations that are used to solve other equations. Furthermore, the same procedures are used to solve formulas that are used to solve the basic equations and literal equations—the algebra axioms of addition, subtraction, multiplication, and division. Letters in a literal equation are handled as though they were numbers in a regular equation. Perform the same operations with literal equations that you performed with the basic equations.

When you have learned how to work with literal equations, you will be able to work with the formulas. This section will help you gain the knowledge and skill you need to work with the formulas.

Remember that the early letters in the alphabet represent known quantities and should be treated as numbers. Study the following examples to learn how to solve literal equations.

Example 7–11

PROBLEM: Find the value of x in the equation $ax = b$.

SOLUTION: Divide both sides of the equation by a.

$$ax = b$$

$$\frac{\cancel{a}x}{\cancel{a}} = \frac{b}{a}$$

$$x = \frac{b}{a}$$

CHECK: Substitute $x = \frac{b}{a}$ into the original equation and evaluate.

$$ax = b$$

$$\cancel{a}\left(\frac{b}{\cancel{a}}\right) = b$$

$$b = b$$

Example 7–12

PROBLEM: Find the value of x in the equation $b(x + 1) = c$.

SOLUTION: Remove the parentheses and transpose $+b$; x will remain on the left side of the equal sign.

$$bx + b = c$$
$$bx = c - b$$

Then divide both sides by b.

$$bx = c - b$$

$$\frac{\cancel{b}x}{\cancel{b}} = \frac{c - b}{b}$$

Thus $x = \frac{c - b}{b}$.

CHECK: Substitute $x = \frac{c - b}{b}$ into the original equation and evaluate.

$$b(x + 1) = c$$

$$b\left(\frac{c - b}{b} + 1\right) = c$$

Perform the operations inside the parentheses first. Note that 1 is changed to $\frac{b}{b}$, which also equals 1.

$$b\left(\frac{c - b}{b} + \frac{b}{b}\right) = c$$

$$b\left(\frac{c - b + b}{b}\right) = c$$

$$\cancel{b}\left(\frac{c}{\cancel{b}}\right) = c$$

$$c = c$$

Example 7–13

PROBLEM: Find the value of y in the equation

$$9y + 12k = 3y + 9k - y$$

SOLUTION: Combine like quantities.

$$9y - 12k = 2y + 9k$$

Transpose the unknown quantities so that all y quantities are on the left side of the equation.

$$9y - 2y = 9k + 12k$$

Combine like quantities.

$$7y = 21k$$

Solve for y by dividing both sides by 7.

$$7y = 21k$$

$$\frac{\cancel{7}y}{\cancel{7}} = \frac{\overset{3}{\cancel{21}}k}{\cancel{7}}$$

$$y = 3k$$

CHECK: Substitute $y = 3k$ into the original equation and evaluate.

$$9y - 12k = 3y + 9k - y$$
$$9(3k) - 12k = 3(3k) + 9k - 3k$$
$$27k - 12k = 9k + 9k - 3k$$
$$15k = 15k$$

Example 7-14

PROBLEM: Find the value of y in the following equation:

$$x(y - 5) + z(y + 6) = x + y$$

SOLUTION: Remove the parentheses.

$$xy - 5x + zy + 6z = x + y$$

Transpose the quantities so that all the y quantities are on the left side and all the other quantities are on the right side and combine.

$$xy - 5x + zy + 6z = x + y$$
$$xy + zy - y = x + 5x - 6z$$
$$xy + zy - y = +6x - 6z$$

Rewrite as follows:

$$y(x + z - 1) = 6(x - z)$$

(This is known as factoring the common quantity from each side of the equation.) Then divide both sides by $(x + z - 1)$.

$$\frac{y\cancel{(x + z - 1)}}{\cancel{(x + z - 1)}} = \frac{6(x - z)}{(x + z - 1)}$$

$$y = \frac{6(x - z)}{(x + z - 1)}$$

CHECK: The checking process is more difficult than working the problem, so check by reworking the complete problem.

EXERCISE 7-3

Solve and check the following exercises. Remember that the beginning letters of the alphabet are considered as known numbers and the later letters of the alphabet as unknowns.

7-46. $6abx = 9abc$

7-47. $k - nx = j$

7-48. $3az = 5az + 2$

7-49. $7x + 8a + 22a = 0$

7–50. $5a^3x = 30$

7–51. $4w = (4a)(8b)$

7–52. $5(p - x) = 3(p - x)$

7–53. $20a = 14a - (x + 2a)$

7–54. $\dfrac{2w}{3} - 4 = 20$

7–55. $7x + 8a - 3x = 3a + 2x$

7–56. $\dfrac{N}{7} + 2 = 8$

7–57. $ax + b - c = 0$

7–58. $8y + (d - 2y) = 4(d - y)$

7–59. $\dfrac{5}{6}(3a - 2x) + \dfrac{1}{6}x = a$

7–60. $(2x - b)(2x + b) + b(b - 8) = 4x(x - 5) - 48b$

THINK TIME

Develop some formulas that could be used in your job. Practice using these formulas to solve current or potential problems. Try to think of areas where formulas may be used if there are no formulas presently available. If you have the opportunity, discuss the formulas with a supervisor. You may find that there are formulas which could be used to solve problems and make work easier and more efficient.

PROCEDURES TO REMEMBER

1. To solve equations by addition and subtraction:
   (a) Remove parentheses, starting with the innermost set of parentheses.
   (b) Combine like quantities on each side of the equation.
   (c) Using either the addition and subtraction axioms or the transposition method, move the unknown quantities to one side of the equation and the known quantities to the other side of the equation.
   (d) Combine like quantities.
   (e) Solve for the unknown.
   (f) Check your solution by substituting the value of the unknown into the original equation and evaluating.
2. To solve equations by multiplication and division:
   (a) Remove parentheses, starting with the innermost set of parentheses.
   (b) Combine like quantities; starting from left to right, perform multiplication and division, then addition and subtraction, from left to right.
   (c) Transpose quantities as necessary so that the positive unknown is on one side of the equal sign and the known quantities on the other.
   (d) Combine like quantities.
   (e) Solve for the unknown by either the multiplication or division axiom.
   (f) Check your solution by substituting the value of the unknown into the original equation and evaluating.
3. To solve literal equations:
   (a) Handle all literal factors as numbers.
   (b) Remove parentheses, starting from the innermost set of parentheses.
   (c) Combine like quantities; starting from the left, perform multiplications and divisions, then addition and substraction from left to right.
   (d) Move the unknown quantity or quantities to one side of the equal sign so that the unknown quantities are positive.
   (e) Combine like quantities.
   (f) Solve for the unknown quantity using either the multiplication or division axiom.
   (g) Make certain that the unknown is a positive value. This may be done by multiplying each side of the equation by -1 or making an interchange of the left and right members.
   (h) Check your solution by substituting the unknown value of the unknown into the original equation and evaluating.

CHAPTER SUMMARY

1. An *equation* is a statement that two algebraic expressions are equal.
2. A *formula* is a rule or procedure referring to a relationship expressed as an equation by letters, symbols, and constant terms.
3. *Literal quantity* means letters or symbols representing numbers.
4. The *left side* of an equation is the quantity to the left of the equal sign, and the *right side* is the quantity to the right of the equal sign.
5. *Addition axiom:* A quantity added to the left side of the equation must also be added to the right side.
6. *Subtraction axiom:* A quantity subtracted from the left side of the equation must also be subtracted from the right side.
7. *Multiplication axiom:* Each side of the equation must be multiplied by the same quantity.
8. *Division axiom:* Each side of the equation must be divided by the same quantity.
9. The solution or the value of the unknown that sastisfies an equation is called the *root* of the equation or the *solution set*.
10. A *literal equation* is an equation in which some or all of the known quantities are represented by letters instead of numbers; frequently, both letters and numbers are included.
11. *Transposition* is the process of moving a quantity from one side of the equal sign to the other. When a quantity is moved from one side of the equation to the other side, the sign of the quantity is changed. This is because equivalent quantities are being added or subtracted from both sides.
12. A *reciprocal* is a quantity which, when multiplied by the quantity, is equal to 1.

CHAPTER TEST

Solve for the unknowns and check.

T–7–1. $x - 7 = 17$

T–7–2. $2x - 3 = 13$

T–7–3. $x + 1 = -8$

T–7–4. $3k + 15 = 5k + 11$

T–7–5. $\dfrac{J}{5} = -20$

T–7–6. $8x = -72 + 168$

T–7–7. $4x - 6 = 5x - 10$

T–7–8. $8y = 35 + y$

T–7–9. $4z - 2(3 + z) = z - 4$

T–7–10. $6n - 8 = 2(2n - 1)$

T–7–11. $y + 3(y - 2) = 2(y + 4)$

T–7–12. $3(x - 4) = 2(x - 2)$

T–7–13. $3x - 2 = 5x + 7$

T–7–14. $\dfrac{1}{2}x + 8 = 15$

Chapter 7 / Simple Equations and Formulas

T-7-15. $9 + \dfrac{x}{8} = 11$

T-7-16. $\dfrac{2}{3}y + \dfrac{1}{6} = \dfrac{1}{2}(y - 8 + 5)$

T-7-17. $4.3y + 2.5 = -0.7y + 22.5$

T-7-18. $y - 2a = 3y + 4a$

T-7-19. $2[(y - 3) - 3(5 - 3y)] = 4(3 + y)$

T-7-20. $3(m - 10) - 2(45 - m) = 0$

T-7-21. $\dfrac{1}{4x} = 16$

T-7-22. $5(a - x) = 3(a - 3x)$

T-7-23. $8(x - 3) - 5(3x + 1) - 6 = 0$

T-7-24. $3(2y - 4) + 2(y - 3) = 5(y + 3)$

T-7-25. $y(y + 4) + 11 = 8(y + 2) + (y - 4)(y - 5)$

8

FORMULA EVALUATION

OBJECTIVE

1. To evaluate simple and complex formulas.

SELF-TEST

This formula evaluation test will determine what you need to study in this chapter. This evaluation covers simple and complex formulas. Evaluate the following formulas using the given values. Be sure to show your work.

S-8-1. If $C = fgh$ and $f = 8, g = 9, h = 10$, find C.

S-8-2. If $y = f^2 + 3g$ and $f = 8, g = 9, h = 10$, find y.

S-8-3. If $m = \dfrac{h}{3} + \sqrt{g}$ and $h = 10, g = 9$, solve for m.

S-8-4. If $k = f^2 + g^2 + h^2 - fg - gh$ and $f = 8, g = 9, h = 10$, find h.

S-8-5. If $p = \dfrac{(fg)^2}{h}$ and $f = 8, g = 9, h = 10$, find h.

S-8-6. When $v = \dfrac{abh}{2}$ and $a = 2, b = 7.5, h = 15\dfrac{1}{4}$, find v.

117

S-8-7. When $V = \pi r^2 h$ and $\pi = 3.14$, $r = 16$, $h = 5$, find V.

S-8-8. When $A = \dfrac{(B + b)a}{2}$ and $B = 10$, $b = 32$, $a = 28$, find A.

S-8-9. Solve $c = \sqrt{a^2 + b^2}$ when $a = 10$ and $b = 24$.

S-8-10. Solve $a = \sqrt{c^2 - b^2}$ when $c = 52$ and $a = 48$.

S-8-11. Solve $H = \dfrac{F + 1}{1.5}$ when $F = 14$.

S-8-12. Solve $D = \sqrt{\dfrac{HP}{6}}$ when $HP = 54$.

S-8-13. Solve $B = \sqrt{(h + a)(h - a)}$ when $h = 30$ and $a = 18$.

S-8-14. Solve $C = \dfrac{5(F - 32)}{9}$ when $F = 95$.

S-8-15. Solve $P = 0.44 d^2 K n$ when $d = \dfrac{1}{2}$, $K = 100$, and $n = 3$.

S-8-16. Solve $L = \dfrac{EFD}{2(D - d)}$ when $E = 0.60$, $F = 130$, $D = 6$, and $d = 4$.

S-8-17. Find $i = prt$ when $p = \$8000$, $r = 8\%$, and $t = 5\dfrac{1}{2}$ years.

S-8-18. Find $A = 0.7854 d^2$ when $d = 0.875$.

S-8-19. Find $V = \frac{4}{3}\pi r^3$ when $\pi = \frac{22}{7}$ and $r = 7$.

S-8-20. Find $A = \sqrt{s(s-a)(s-b)(s-c)}$ if $s = \frac{1}{2}(a + b + c)$, $a = 13$, $b = 11$, and $c = 16$.

INTRODUCTION

Perhaps like many students you wonder, "How will I ever use algebra once I finish school?" Strange as it may seem, algebra is a tool of industry. In fact, formulas can be traced back to application of early arithmetic.

Consider the problem posed to an employee of the electric company. She was given the task of building a fence around an electric power substation. To find the amount of fencing she would need, she used a formula learned as a child. The formula: $P = 2l + 2w$. In other words, the distance around the substation, or the perimeter (P), is equal to 2 times the length ($2l$) plus 2 times the width ($2w$). Knowing this formula saved time—she did not need to make all of the measurements. By measuring one side and the width, she could use the formulas to determine the perimeter.

Formulas save time and money. They are essential to the world of work. Formulas are algebra.

DEFINITIONS

In Chapter 7 a definition of a formula was given. Here is another way to consider how a formula may be defined. A *formula* uses mathematical language to express the relationship between two or more variables. For example, the formula for finding the number of miles a car will travel on 1 gallon of gasoline is

$$\text{Miles per gallon} = \frac{\text{miles traveled}}{\text{gallons of gasoline used}}$$
$$= \frac{m}{g}$$

Evaluating a formula is not difficult. It is important to have the formula in proper form and follow given procedures.

Most students who have difficulty with mathematics in general and formula evaluation in particular have difficulty because they fail to read the problem correctly. Thus they are unable to perform the designated operations. This is shown by skipping essential steps, failing to show the work, and a lack of understanding of what answer is to be found.

As you complete the following pages, carefully study the formulas given in the examples, noting the steps involved in reaching a solution. As you are given an opportunity to work with formulas, follow the same steps as in the examples. Be certain to show your work, and check the examples when necessary.

EVALUATING FORMULAS USING ADDITION OR SUBTRACTION

The following are examples of formulas that use the operations of addition and subtraction.

Example 8–1

Perimeter of a Triangle

$P = a + b + c$

The formula for the perimeter (distance around the outside) of a triangle is $P = a + b + c$. This means that the perimeter of a triangle is equal to side a plus side b plus side c. See Figure 8–1.

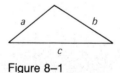

Figure 8–1

PROBLEM: What is the perimeter of a triangle when $a = 5$, $b = 6$, and $c = 9$?

SOLUTION:

Write the formula: $P = a + b + c$
Substitute the givens: $P = 5 + 6 + 9$
Evaluate: $P = 20$

We call this process the *evaluation* of a formula.

Chapter 8 / Formula Evaluation   119

Example 8–2

PROBLEM: The formula for a certain kind of friction is $K = x + y - z$. Evaluate the formula if $x = 3$, $y = 7$, and $z = 4$.

SOLUTION:

Write the formula:	$K = x + y - z.$
Substitute the givens:	$K = 3 + 7 - 4$
Evaluate:	$K = 6$

EXERCISE 8–1

Evaluate the following formulas using addition and subtraction.

8–1. Find the perimeter of the triangle shown in Figure 8–2.

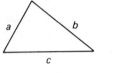

a
b
c
a = 5
b = 7
c = 8
P = a + b + c

Figure 8–2

8–2. Find the perimeter of the trapezoid shown in Figure 8–3.

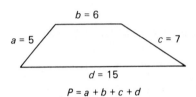

b = 6
a = 5
c = 7
d = 15
P = a + b + c + d

Figure 8–3

8–3. Find the perimeter of a 6-inch square. See Figure 8–4.

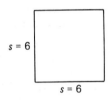

s = 6
s = 6
P = 4s

Figure 8–4

8–4. Find the missing dimension x in Figure 8–5.

39
25
x

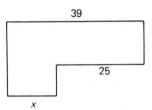

Figure 8–5

8–5. Find the missing dimension y in Figure 8–6.

40
11
7
y

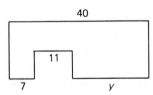

Figure 8–6

8–6. Solve the formula $K = x + y - z$ for y when $K = 20$, $x = 5$, and $z = 4$.

8–7. Find the perimeter of the property in Figure 8–7.

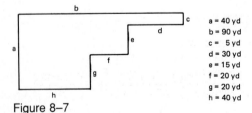

b
c
d
e
a
f
g
h
a = 40 yd
b = 90 yd
c = 5 yd
d = 30 yd
e = 15 yd
f = 20 yd
g = 20 yd
h = 40 yd

Figure 8–7

8–8. Using the formula for determining profit, $p = g - c - l - j$, find the cost of manufacturing the products when $p = \$20,000$, $g = \$57,000$, $l = \$10,000$, and $j = \$1500$.

8-9. Using a formula for balancing a checkbook, $B = A - a - b - c$, find the amount of check a when $B = \$99$, $A = \$257$, $b = \$77$, and $c = \$19$.

8-10. Using a formula for finding caloric intake, $C = a + b + c + d + e$, find the value of e when $C = 401$, $a = 75$, $b = 110$, $c = 130$, and $d = 22$.

EVALUATING FORMULAS USING MULTIPLICATION

The formulas in the preceding section used addition and subtraction. The formulas in this section make use of multiplication.

Example 8–3

Area of a Rectangle
The formula for finding the area of a rectangle is $A = lw$. This means that the area (A) of a rectangle is equal to the length (l) times the width (w).

PROBLEM: Find the area of a rectangle when the length (l) is 10 feet and the width (w) is 8 feet.

SOLUTION:
    Write the formula: $A = lw$
    Substitute the givens: $A = (10 \text{ ft})(8 \text{ ft})$
    Evaluate: $A = 80$ sq ft

Example 8–4

Area of a Square
The formula for finding the area of a square is $A = s^2$. This means that the area (A) of a square is equal to the length of one side (s) times the length of another side (s), or $A = (s)(s)$ or s^2.

PROBLEM: Find the area of a square when $s = 4$ feet.

SOLUTION:
    Write the formula: $A = s^2$
    Substitute the givens: $A = (4 \text{ ft})(4 \text{ ft})$
    Evaluate: $A = 16$ sq ft

Example 8–5

Circumference of a Circle
The formula for finding the circumference (distance around) a circle is $C = d\pi$. This means that the circumference (C) is equal to the diameter (d) times π, or 3.142.

PROBLEM: Find the circumference of a circle when $d = 7$ feet.

SOLUTION:
    Write the formula: $C = d\pi$
    Substitute the givens: $C = (7 \text{ ft})(3.142)$
    Evaluate: $C = 22$ ft

Example 8–6

Distance
The formula for finding distance is $d = rt$. This means that distance (d) equals the rate (r) times the time (t).

PROBLEM: Find distance traveled when $r = 55$ miles per hour and $t = 240$ minutes.

SOLUTION:
Write the formula: $d = rt$

Substitute the givens:

$$d = \left(\frac{55 \text{ miles}}{\text{hour}}\right)\left(\frac{1 \text{ hour}}{60 \text{ minutes}}\right)\left(\frac{\overset{4}{\cancel{240 \text{ minutes}}}}{1}\right)$$

Evaluate: $d = 220$ miles

Note: Only like units may be multiplied. Convert to like units when necessary.

Example 8–7

Interest

The formula for finding interest is $i = prt$. This means that the amount of interest (i) is equal to the principal (p) times the rate of interest (r) times the time in years (t).

PROBLEM: Find the interest when $p = \$5000$, $r = 0.08$, and $t = 3$ years.

SOLUTION:
Write the formula: $i = prt$
Substitute the givens: $i = (\$5000)(0.08)(3)$
Evaluate: $i = \$1200$

EXERCISE 8–2

Solve the following problems, performing the operations indicated.

8–11. Find the area of a rectangle that is 15 inches by 11 inches.

$$A = lw \quad l = 15 \text{ in.} \quad w = 11 \text{ in.}$$

8–12. Find the distance traveled if the rate is 40 miles per hour and the time is 20 hours.

$$d = rt \quad r = 40 \text{ mph} \quad t = 20 \text{ hr}$$

8–13. Find the interest if the principal is $1000, the rate is 7%, and the time is 25 years.

$$i = prt \quad r = 0.07 \quad t = 25 \text{ years}$$

8–14. Find the volume of a rectangular solid that is 30 feet by 12 feet by 8 inches.

$$V = lwh \quad l = 30 \text{ ft} \quad w = 12 \text{ ft} \quad h = 8 \text{ in.}$$

8–15. Find the circumference of a circle 8 yards in diameter.

$$C = d\pi \quad d = 8 \text{ yd} \quad \pi = 3.1416$$

8–16. Convert 27 feet to inches. (1 foot = 12 inches.)

8–17. Find the circumference of a circle if the radius is 34 inches.

$$C = 2\pi r \qquad r = 34 \text{ in.} \qquad \pi = 3.14$$

8–18. Find the volume of a rectangular solid with a 36-square-foot base and a height of 24 inches. *Remember to convert inches to feet.* (12 inches = 1 foot.)

$$V = Bh \qquad B = 36 \text{ sq ft} \qquad h = 24 \text{ in.}$$

8–19. Find the perimeter of a rectangle 7 feet long and 3 feet wide.

$$p = 2l + 2w \qquad l = 7 \text{ ft} \qquad w = 3 \text{ ft}$$

8–20. Find the area of a 41-foot square.

$$A = s^2 \qquad s = 41 \text{ ft}$$

8–21. Find the circumference in *feet* of a circle when the radius is 28.5 inches.

$$C = 2\pi r \qquad r = 28.5 \text{ in.} \qquad \pi = 3.142$$

8–22. Find the surface area of a 3-foot cube.

$$S = 6e^2 \qquad e = 3 \text{ ft}$$

8–23. Find the surface area of a rectangular solid 6 feet long by 5 feet wide and 3 feet high.

$$S = 2lw + 2hl + 2hw \qquad l = 6 \text{ ft} \qquad w = 5 \text{ ft}$$
$$h = 3 \text{ ft}$$

8–24. Find the simple interest for $298 at 5.5% for 3 years.

$$i = prt \qquad p = \$298 \qquad r = 5.5\% \qquad t = 3 \text{ years}$$

8–25. Find the diameter, to the nearest tenth of a foot, of a tree with a circumference of 282.75 inches. *Remember to convert inches to feet.* (12 inches = 1 foot.)

$$d = \frac{C}{\pi} \qquad C = 282.75 \text{ in.}$$

Chapter 8 / Formula Evaluation

EVALUATING FORMULAS USING DIVISION OR ROOTS

The formulas included thus far required using addition, subtraction, and multiplication. More difficult formulas may include operations of division and roots. Study the following examples, making special note of the procedures involved.

Example 8–8

Area of a Triangle

The formula for determining the area of a triangle is $A = bh/2$ or $\frac{1}{2}bh$ or $0.5bh$. This means that the area (A) of the triangle is equal to the base (b) times the height (h) divided by 2.

PROBLEM: Find the area of a triangle when the base is $b = 4$ feet, and the height is $h = 12$ feet. See Figure 8–8

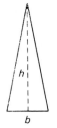

Figure 8–8

SOLUTION:

Write the formula: $A = \dfrac{bh}{2}$

Substitute the givens: $A = \dfrac{(4 \text{ ft})(12 \text{ ft})}{2}$

Evaluate: $A = 24$ sq ft

Example 8–9

Area of a Trapezoid

The formula for finding the area of a trapezoid is $A = \frac{1}{2}(b_1 + b_2)h$. This means that the area (A) of the trapezoid is equal to one half the sum of the base 1 (b_1 is read "b sub 1") and base 2 (b_2) times the height (h).

PROBLEM: Find the area of the trapezoid: $b_1 = $ 10 inches, $b_2 = 20$ inches, and $h = 5$ inches. See Figure 8–9

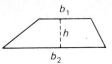

Figure 8–9

SOLUTION:

Write the formula: $A = \dfrac{1}{2}(b_1 + b_2)h$

Substitute the givens: $A = \dfrac{1}{2}(10 \text{ in.} + 20 \text{ in.})(5 \text{ in.})$

$A = 0.5(30 \text{ in.})(5 \text{ in.})$

Evaluate: $A = 75$ sq in.

Example 8–10

Depth

The formula for finding the depth of a certain type of thread is

$$D = d - \dfrac{1.732}{N}$$

This means that the depth of the thread (D) is equal to the depth of one thread (d) minus 1.732 divided by the number of threads (N).

PROBLEM: Find the depth of the thread when $d = 0.75$ and $N = 5$.

SOLUTION:

Write the formula: $D = d - \dfrac{1.732}{N}$

Substitute the givens: $D = 0.075 - \dfrac{1.732}{5}$

Evaluate: $D = 0.75 - 0.3464$
$= 0.4036$

Example 8–11

Height of Isosceles Triangle

The formula for finding the height of an isosceles triangle given the equal sides and base is

$$h = \sqrt{a^2 - \dfrac{b^2}{4}}$$

This means that the height (h) of the isoceles triangle is equal to the square root ($\sqrt{}$) of one of the equal sides squared (a^2) minus the base squared divided by 4, written as $\dfrac{b^2}{4}$

PROBLEM: Find the height of an isoceles triangle when $a = 26$ feet and $b = 20$ feet. Use a calculator to find the square root.

SOLUTION:

Write the formula: $h = \sqrt{a^2 - \dfrac{b^2}{4}}$

Substitute the givens: $h = \sqrt{(26)^2 - \dfrac{(20)^2}{4}}$

Evaluate:

$h = \sqrt{676 - \dfrac{400}{4}}$

$h = \sqrt{676 - 100}$

$h = \sqrt{576}$

$h = 24$ ft

EXERCISE 8–3

Solve the following problems, make certain to perform the indicated operations. Remember to convert measurements to the same unit.

8–26. $A = \dfrac{1}{2}(b_1 + b_2)h$.
Find A when $b_1 = 2$ feet, $b_2 = 36$ inches, and $h = 24$ inches.

8–27. $E = \dfrac{360}{n}$.
Find E when $n = 4$.

8–28. $A = \dfrac{0.7854nd^2}{360°}$.
Find A when $n = 60°$ and $d = 12$ inches.

8–29. $K = \dfrac{L(D - d)}{l}$.
Find K when $L = 3.5, D = 0.5, d = 0.25,$ and $l = 0.25$.

8–30. $M = (D - 1.732P) + 3W$.
Find M when $D = 7.5, P = 10,$ and $W = 0.06$.

8–31. $A = 0.7854d^2$.
Find A when $d = 11$ feet.

8–32. $S = 2\pi r(r + h)$.
Find S when $\pi = \dfrac{22}{7}, r = 2$ feet, and $h = 1\dfrac{1}{2}$ feet.

8–33. $C = \dfrac{5}{9}(F - 32°)$.
Find C when $F = 131°$

Chapter 8 / Formula Evaluation

8-34. $V = \dfrac{4\pi r^3}{3}$.
Find V when $r = 6$ yards and $\pi = 3.14$.

8-35. $F = \dfrac{9}{5}C + 32°$.
Find F when $C = 115°$.

8-36. $W = \dfrac{Kbd^2}{L}$.
Find W when $K = 6000$, $b = 6$, $d = 8$, and $L = 24$.

8-37. $S = \dfrac{gt^2}{2}$.
Find S when $g = 32$ and $t = 4$.

8-38. $S = \dfrac{n}{2}(a + l)$.
Find S when $a = 10$, $l = 6$, and $n = 50$.

8-39. $H = \dfrac{DV}{375}$.
Find H when $D = 187.5$ and $V = 200$.

8-40. $K = 2a - 5(n - 1)$.
Find K when $a = 8$ and $n = 3$.

8-41. $R = \dfrac{1}{\dfrac{1}{r_1} + \dfrac{1}{r_2}}$.
Find R when $r_1 = 3$ and $r_2 = 4$.

8-42. $A = \dfrac{D}{0.5(D - d)}$.
Find A when $D = 10$ and $d = 4$.

8-43. $A = 3.1416(R + r)(R - r)$.
Find A when $R = \dfrac{3}{2}$ and $r = \dfrac{3}{8}$.

8–44. $E = \dfrac{(N)P}{NP + R}$.
Find E when $N = 2$, $P = 4$, and $R = 5.3$.

8–45. $L = \sqrt{b^2 + h^2}$.
Find L when $b = 4.75$ and $h = 3.625$.

8–46. $b = \sqrt{h^2 - a^2}$.
Find b when $h = 10\dfrac{1}{2}$ and $a = 8\dfrac{3}{4}$.
(*Hint:* Convert $10\dfrac{1}{2}$ to 10.5 and $8\dfrac{3}{4}$ to 8.75.)

8–47. $d = \sqrt{\dfrac{4A}{\pi}}$.
Find d when $A = 490.875$ and $\pi = 3.14159$.

8–48. $r = \sqrt{\dfrac{A}{\pi}}$.
Find r when $A = 28.26$ and $\pi = 3.14$.

8–49. $l = 2\sqrt{h(2r - h)}$.
Find l when $h = 6$ and $r = 15$.

8–50. $A = \sqrt{s(s - a)(s - b)(s - c)}$.
Find A when $s = \dfrac{1}{2}(a + b + c)$ and $a = 8$, $b = 6$, and $c = 4$.

The exercises you have completed have involved formula evaluation. In the next chapter we discuss how to manipulate or transform formulas into the desired forms.

THINK TIME

Write four formulas used in your regular daily living. Using these formulas, with the knowledge and skills learned in this chapter, solve for the unknowns. Think of additional situations in which you could use these formulas. Practice solving problems with the stated formulas.

PROCEDURES TO REMEMBER

1. To evaluate formulas:
   (a) Read the formula to determine its purpose.
   (b) Reread the formula, making certain of the given quantities and the desired quantity.
   (c) Substitute the given quantities into the formula, making certain that you have used the correct dimensions.
   (d) Estimate your answers so that you will have a reasonable idea of the possible solution.
   (e) Evaluate; show all necessary mathematical operations to prevent errors.
   (f) Be sure that your solution is written in the correct dimensions.

CHAPTER SUMMARY

A formula uses mathematical language to express the relationship between two or more variables. Thus a formula is a mathematical abbreviation for an accepted operational procedure. Formula evaluation is the process of obtaining the value of a formula from the given variables.

CHAPTER TEST

Evaluate the following using the given quantities.

T-8-1. When $x = 5$, $y = 3$, and $z = 2$, $\dfrac{x + y}{z} = ?$

T-8-2. When $a = 15$, $b = 14$, and $c = 12$, $a^2 + b^2 - c^2 = ?$

T-8-3. When $x = 1$, $y = 3$, and $z = 2$, $x^2 + y^2 + z^3 = ?$

T-8-4. When $w = -2$ and $k = 3$, $\dfrac{2w + 5k}{7} = ?$

T-8-5. When $x = 1$, $y = 3$, $z = 2$, and $w = -2$, $xy + yz + zw = ?$

T-8-6. Evaluate $\dfrac{1}{x} + \dfrac{3}{y}$ when $x = 1$ and $y = -1$.

T-8-7. Evaluate $(x + z)(w + k)$ when $x = 1$, $z = 2$, $w = -2$, and $k = 3$.

T-8-8. Evaluate $(ab)^2(c - b)^2$ when $a = 4$, $b = 0.5$, and $c = 2.5$.

T-8-9. Evaluate $(a + b)^2(a - b)^2$ when $a = 6$ and $b = 4$.

T-8-10. Evaluate $(x + y - z)^3 = $ when $x = 5$, $y = 12$, and $z = 7$.

T-8-11. The formula for the area of an ellipse is $A = \pi ab$. Find A when $\pi = 3.14$, $a = 20$, and $b = 12$.

T-8-12. The formula for the perimeter of an ellipse is $P = \pi(a + b)$. Find P when $\pi = 3.14$, $a = 24$, and $b = 21$.

T-8-13. If $l = 2\sqrt{h(2r - h)}$, find l when $h = 4$ and $r = 10$. This is the formula for finding the length of a segment when you know the radius of the circle and the height of the segment.

T-8-14. The formula for the area of a ring is $A = \pi(R + r)(R - r)$. Find A when $R = \frac{7}{8}$, $r = \frac{3}{4}$, and $\pi = 3.14$. (*Hint:* Convert the fractions to decimals.)

T-8-15. The formula for the radius of a circle, given the area, is $r = \sqrt{\dfrac{A}{\pi}}$. Find r when $\pi = 3.14$ and $A = 502.4$.

T-8-16. The formula for the volume of a sphere is $V = \dfrac{4\pi r^3}{3}$. Find the volume of a sphere (in cubic yards) when the radius is 12 feet.

T-8-17. The formula for the volume of a cylinder is $V = \pi r^2 h$. Find the volume of a cylinder when $\pi = 3.14$, $r = 5$, and $h = 12$.

T-8-18. The formula for the area of a sector of a circle is $A = \dfrac{\theta}{360}\pi r^2$. Find A when $\theta = 75$, $\pi = 3.14$, and $r = 8$.

T-8-19. The formula for the diameter of a circle given the area is $d = \sqrt{A \div \dfrac{\pi}{4}}$. Find d when $A = 12.56$ and $\pi = 3.14$.

T-8-20. The formula for the area of a triangle, given all the sides, is
$$A = \sqrt{s(s - a)(s - b)(s - c)}$$
$$s = \frac{1}{2}(a + b + c)$$

Find A when $a = 28$, $b = 25$, and $c = 17$.

Chapter 8 / Formula Evaluation

9

FORMULA TRANSPOSITION

OBJECTIVES

1. To solve formulas for the desired unknown.

SELF-TEST

This test on formula transposition will determine what you need to study in this chapter. If you are able to solve the formulas, you are ready to move to the next chapter. Be sure to show your work so that you can determine where you made errors and what you need to do to correct your mistakes.

S-9-1. Solve $P = a + b + c$ for c.

S-9-2. Solve $V = \frac{1}{3} Bh$ for B.

S-9-3. Solve $L = 2\pi rh$ for h.

S-9-4. Solve $P = \frac{w}{t}$ for t.

S-9-5. Solve $HP = \frac{fl}{t}$ for t.

S-9-6. Solve $H = \frac{D^2 N}{2.5}$ for D.

S-9-7. Solve $V = \frac{1}{3} \pi r^2 h$ for h.

S-9-8. Solve $A = \frac{h}{2}(b_1 + b_2)$ for b_2.

S–9–9. Solve $S = 2\pi rh + 2\pi r^2$ for h.

S–9–10. Solve $V = \pi r^2 h$ for r.

S–9–11. Solve $V = \dfrac{4}{3}\pi r^3$ for r.

S–9–12. Solve $DS = ds$ for d.

S–9–13. Solve $S = \dfrac{CD}{12d}$ for D.

S–9–14. Solve $W = \dfrac{5bd^2K}{2L}$ for K.

S–9–15. Find a when $T = 384$ and $T = 6a^2$.

S–9–16. Find r when $A = \dfrac{5\pi}{16}$, $R = \dfrac{3}{4}$, and $A = \pi R^2 - \pi r^2$.

S–9–17. Find b when $P = 51\pi$, $a = 32$, and $P = \pi(a + b)$.

S–9–18. Find r when $V = \pi R^2 h - \pi r^2 h$, $R = 8$, $h = 10$, $V = 77.5\pi$, and $\pi = 3.14$.

S–9–19. Solve $S = 4\pi r^2$ for r; then find r when $S = 484\pi$.

S–9–20. Find d if $V = \dfrac{32\pi}{3}$ and $V = \dfrac{\pi d^3}{6}$.

S-9-21. Solve $A = \dfrac{\theta \pi r^2}{360}$ for r.

S-9-22. Solve $r = \dfrac{\left(\dfrac{w}{2}\right)^2 + h^2}{2h}$ for w.

S-9-23. Solve $V = \dfrac{\pi h}{3}(a^2 + ab + b^2)$ for h, then find h when $a = 24$, $b = 18$, and $V = 3552$.

S-9-24. Solve $P = \pi\sqrt{2(a^2 + b^2)}$ for b.

S-9-25. Solve $\dfrac{1}{R} = \dfrac{1}{r_1} + \dfrac{1}{r_2} + \dfrac{1}{r_3}$ for r_3 and find r_3 when $R = \dfrac{12}{13}$, $r_1 = 2$, and $r_2 = 3$.

INTRODUCTION

Given formulas or formulas found in handbooks are not always in the desired form. Therefore, it is important you know how to solve formulas so that they may work for you. This is not a difficult task if you work through the process step by step. Many students find solving formulas or transposition of formulas very difficult because they fail to read the formulas correctly. For example, the formula $A = \dfrac{1}{2}(b_1 + b_2)h$ could be misread as $A = \dfrac{1}{2}b_1 = 2 \times h$. Failure to complete work inside the parentheses first will result in an incorrect answer. Correct reading of formulas can help to prevent confusion and error.

SOLVING FORMULAS USING INVERSE OPERATIONS

Reading the original formula correctly becomes very important when you solve formulas. When the formula is read correctly, you will be able to proceed to solve the formula correctly. The solutions are often made through inverse operations. This means that division is used to solve an equation containing multiplication. Study the following examples and note how inverse operations are used to solve the equations.

Example 9-1

> **Area of a Rectangle**
> The formula for the area of a rectangle equals the length times the width, or $A = lw$.

PROBLEM: Solve the formula for w.

SOLUTION: Remember, division is the inverse operation of multiplication. Thus divide both sides by l to find w.

$$A = lw$$
$$\dfrac{A}{l} = \dfrac{\cancel{l}w}{\cancel{l}}$$
$$\dfrac{A}{l} = w$$

The width (w) equals the area (A) divided by the length (l).

$$w = \frac{A}{l}$$

Example 9–2

PROBLEM: Find w for Example 9–1 when $A = 44$ square feet and $l = 11$ feet.

SOLUTION:

Write the formula: $\quad w = \dfrac{A}{l}$

Substitute the givens: $\quad w = \dfrac{\overset{4}{\cancel{44}} \text{ sq ft}}{\underset{1}{\cancel{11}} \text{ ft}}$

Evaluate: $\quad w = 4$ ft

Example 9–3

This example involves the operations of multiplication and addition.

Perimeter of a Rectangle
The perimeter of a triangle equals 2 times the width plus 2 times the length, or

$$P = 2l + 2w$$

PROBLEM: Solve the formula for w.

SOLUTION: Isolate the quantity w on one side of the equation by subtracting $2l$ from each side of the equation and combining like quantities.

$$P = 2l + 2w$$
$$P - 2l = 2l + 2w - 2l$$
$$P - 2l = 2w$$

Solve the equation for w by dividing each side of the equation by 2.

$$\frac{\overset{1}{\cancel{2}}w}{\underset{1}{\cancel{2}}} = \frac{P - 2l}{2}$$

$$w = \frac{P - 2l}{2}$$

Example 9–4

PROBLEM: Find w for Example 9–3 when $P = 38$ and $l = 12$.

SOLUTION:

Write the formula: $\quad w = \dfrac{P - 2l}{2}$

Substitute the givens: $\quad w = \dfrac{38 - 2(12)}{2}$

$$w = \frac{38 - 24}{2}$$

Evaluate: $\quad w = \dfrac{\overset{7}{\cancel{14}}}{\underset{1}{\cancel{2}}}$

Example 9–5

This example involves the operations of multiplication and division.

Area of a Triangle
The area of a triangle equals one half the base times the height, or $A = \dfrac{1}{2} bh$.

PROBLEM: Solve the formula for h.

SOLUTION: When a formula involves a fraction, change the fraction to a whole number by multiplication. First find the lowest common denominator, which is 2. Then multiply each term on both sides of the equation by 2.

$$A = \frac{1}{2} bh \quad \text{or} \quad A = \frac{bh}{2}$$

$$2A = \cancel{(2)} \frac{1}{\underset{1}{\cancel{2}}} bh$$

Then to solve for h, divide each side of the equation by b.

$$2A = bh$$

$$\frac{2A}{b} = \frac{\overset{1}{\cancel{b}}h}{\underset{1}{\cancel{b}}}$$

$$h = \frac{2A}{b}$$

Example 9–6

PROBLEM: Find b for Example 9–5 when $A = 360$ square feet and $b = 40$ feet.

SOLUTION:

Write the formula: $\quad h = \dfrac{2A}{b}$

Chapter 9 / Formula Transposition

Substitute the givens: $h = \dfrac{2(360 \text{ sq ft})}{40 \text{ ft}}$

Evaluate: $h = \dfrac{\cancel{720} \text{ sq ft}^{18 \text{ ft}}}{\cancel{40 \text{ ft}}_{1}} = 18 \text{ ft}$

EXERCISE 9–1

Solve the following problems. Make sure that you read them carefully and solve for the correct letter. Show all necessary work.

9–1. Solve $Q = \dfrac{WL}{T}$ for T. Find T when $Q = 250$, $W = 125$, and $L = 20$.

9–2. Solve $i = prt$ for t. Find t when $p = \$500$, $r = 0.08$, and $i = \$120$.

9–3. Solve $C = 2\pi r$ for r. Find r when $C = 154$ and $\pi = \dfrac{22}{7}$.

9–4. Solve $p = a + b + c$ for c. Find c when $p = 100$, $a = 21$, and $b = 17$.

9–5. Solve $V = Bh$ for B. Find B when $V = 1000$ cubic feet and $h = 8$ feet.

9–6. Solve $Q = 0.000477EIT$ for T. Find T when $Q = 954$, $E = 1000$, and $I = 20$.

9–7. Solve $C = \dfrac{5}{9}(F - 32°)$ for F. Find F if $C = 100°$.

9–8. Solve $x - y = xy$. Find y when $x = 8$.

9–9. Solve $S = \pi rs$ for s. Find s when $S = 4352$, $\pi = 3.14$, and $r = 462$.

9–10. Solve $P = 2a + b$ for a. Find a when $P = 28$ and $b = 12$.

9-11. Solve $i = prt$ for r. Find r when $i = \$80$, $p = \$500$, and $t = 2$ years.

9-12. Solve $C = \pi d$ for d. Find d when $C = 120$ feet and $\pi = 3.14$, to the nearest hundredth.

9-13. Solve $L = \dfrac{\pi r A}{180}$ for A. Find A when $L = 314$, $r = 60$, and $\pi = 3.14$.

9-14. Solve $T = \dfrac{1}{2} ps + B$ for B. Find B when $T = 580$, $p = 58$, and $s = 12$.

9-15. Solve $T = 2B + pH$ for H. Find H when $T = 860$, $B = 45$, and $p = 11$.

SOLVING FORMULAS USING ROOTS

The procedures of transposition become more difficult when finding roots. You may wish to review what you have learned about the extraction of roots in previous chapters and what you have learned about transposition in this chapter. Carefully study the following examples.

Example 9-7

> **Surface Area of a Cube**
> The surface area of a cube equals 6 times the length of one edge square, or $A = 6e^2$.

FORMULA: The surface area of a cube equals 6 times the length of one edge squared: $A = 6e^2$.

PROBLEM: Solve the formula for e.

SOLUTION: Divide both sides by 6 to isolate e^2 on one side of the equation.

$$A = 6e^2$$

$$\frac{A}{6} = \frac{\cancel{6}e^2}{\cancel{6}}$$

$$\frac{A}{6} = e^2$$

To solve for e, take the square root of each side of the equation.

$$e^2 = \frac{A}{6}$$

$$\sqrt{e^2} = \sqrt{\frac{A}{6}}$$

$$e = \sqrt{\frac{A}{6}}$$

Example 9-8

PROBLEM: Find e for Example 9-7 if $A = 216$ square feet.

SOLUTION:

Write the formula $\quad e = \sqrt{\dfrac{A}{6}}$

Substitute the givens: $\quad e = \sqrt{\dfrac{\cancel{216}^{36} \text{ sq ft}}{\cancel{6}}}$

$$e = \sqrt{36 \text{ sq ft}}$$

Evaluate: $\quad e = 6$ ft

Chapter 9 / Formula Transposition

Example 9-9

> **Area of a Circle**
> The area of a circle equals π (approximately 3.14) times the radius of the circle squared, or $A = \pi r^2$.

PROBLEM: Solve the formula for r.

SOLUTION: Divide both sides by π to isolate r^2 on one side of the equation.

$$A = \pi r^2$$

$$\frac{A}{\pi} = \frac{\cancel{\pi} r^2}{\cancel{\pi}}$$

$$\frac{A}{\pi} = r^2$$

Solve the formula for r by taking the square root of each side of the equation.

$$r^2 = \frac{A}{\pi}$$

$$\sqrt{r^2} = \sqrt{\frac{A}{\pi}}$$

$$r = \sqrt{\frac{A}{\pi}}$$

Example 9-10

PROBLEM: Find r for Example 9-9 if the area of a circle is 50.24 and $\pi = 3.14$.

SOLUTION:

Write the formula: $r = \sqrt{\dfrac{A}{\pi}}$

Substitute the givens: $r = \sqrt{\dfrac{50.24 \text{ sq yd}}{3.14}}$

Evaluate: $r = \sqrt{16 \text{ sq yd}}$

$r = 4 \text{ yd}$

Example 9-11

> **Volume of a Sphere**
> The volume of a sphere equals 4 times π times the radius of the sphere cubed divided by 3, or
> $$V = \frac{4\pi r^3}{3}$$

PROBLEM: Solve the formula for r.

SOLUTION: Note that r is cubed; thus we will need to find the cube root of r^3. First, clear the fraction, $\dfrac{4\pi r^3}{3}$, by multiplying both sides of the equation by 3.

$$V = \frac{4\pi r^3}{3}$$

$$(3)V = \frac{\cancel{(3)}^1 \, 4\pi r^3}{\cancel{3}_1}$$

$$3V = 4\pi r^3$$

Isolate r^3 by dividing both sides of the equation by 4π.

$$3V = 4\pi r^3$$

$$\frac{3V}{4\pi} = \frac{\cancel{4\pi} r^3}{\cancel{4\pi}}$$

$$\frac{3V}{4\pi} = r^3$$

Solve the formula for r, by taking the cube root of each side of the equation.

Note: Upscale calculators will give the cube root by using the $\boxed{x^y}$ key.

$$r^3 = \frac{3V}{4\pi}$$

$$\sqrt[3]{r^3} = \sqrt[3]{\frac{3V}{4\pi}}$$

$$r = \frac{\sqrt[3]{3V}}{4\pi}$$

Example 9-12

PROBLEM: Find r for Example 9-11 if $V = 523.333$ cubic units and $\pi = 3.14$.

SOLUTION:

$$r = \sqrt[3]{\frac{3V}{4\pi}}$$

$$r = \sqrt[3]{\frac{(3)(523.333) \text{ m}^3}{4(3.14)}}$$

$$r = \sqrt[3]{\frac{1570 \text{ m}^3}{12.56}}$$

$$r = \sqrt[3]{125 \text{ m}^3}$$

$$r = 5 \text{ m}$$

Note: To obtain the cube root with a calculator, key in 125, then press the $\boxed{x^y}$ key; then $.\overline{3}$ (which is the $\frac{1}{3}$ power) and press the $\boxed{=}$ key.

Example 9–13

Height of Isosceles Triangle

The height of an isosceles triangle is equal to the square root of one of the equal sides squared minus the base squared divided by 4, or

$$h = \sqrt{a^2 - \frac{b^2}{4}}$$

PROBLEM: Solve the formula for a.

SOLUTION: To find the value of a, square both sides of the equation.

$$(h)^2 = \left(\sqrt{a^2 - \frac{b^2}{4}}\right)^2$$

$$h^2 = a^2 - \frac{b^2}{4}$$

Then add $\frac{b^2}{4}$ to both sides of the equation.

$$\frac{b^2}{4} + h^2 = a^2 - \frac{b^2}{4} + \frac{b^2}{4}$$

$$\frac{b^2}{4} + h^2 = a^2$$

Take the square root of each side.

$$\sqrt{a^2} = \sqrt{\frac{b^2}{4} + h^2}$$

$$\text{Thus } a = \sqrt{\frac{b^2}{4} + h^2}$$

Example 9–14

PROBLEM: Find the value of a in Example 9–13 when $h = 24$ and $b = 20$.

SOLUTION:

Write the formula: $a = \sqrt{\dfrac{b^2}{4} + h^2}$

Substitute the givens: $a = \sqrt{\dfrac{(20)^2}{4} + (24)^2}$

Evaluate: $a = \sqrt{\dfrac{400}{4} + 576}$

$a = \sqrt{676}$

$a = 26$ units

EXERCISE 9–2

Solve the following problems. Study each problem carefully to make sure that you understand for what quantity the problem is to be solved. If you have difficulty, study the examples.

9–16. Solve $A = \dfrac{\pi r^2}{2}$ for r. Find r when $A = 157$ square units and $\pi = 3.14$.

9–17. Solve $x = ky^2$ for y. Find y when $x = 128$ and $k = 8$.

9–18. Solve $F = \dfrac{kM}{d^2}$ for d. Find d when $k = 28$, $M = 8$, and $F = 14$.

9–19. Solve $S = \dfrac{at^2}{2}$ for t. Find t when $a = 4$ and $S = 128$.

9–20. Solve $F = \dfrac{kmM}{d^2}$ for M. Find M when $d = 3$, $F = 42$, $k = 14$, and $m = \dfrac{1}{3}$.

9–21. Solve the mass-energy equation $e = mc^2$ for c. Find c if $e = 6348$ and $m = 12$.

9–22. Solve $V = e^3$ for e. This is the formula for the volume of a cube. Find e when $V = 729$ cubic yards.

9–23. Solve $S = 4\pi r^2$, which is the formula for the surface area of a sphere, for r. Find r when $S = 784\pi$ square feet.

9–24. Solve the volume of a sphere formula, $V = \dfrac{\pi d^3}{6}$, for d. Find d when $V = \dfrac{171.5\pi}{3}$.

9–25. Solve $r = \sqrt{\dfrac{A}{\pi}}$ for A. This is the formula for the radius of a circle when you know the area. Find A when $r = 7$ and $\pi = \dfrac{22}{7}$.

9–26. Solve the algebra equation $(y - a)^2 - b = 4ab$ for $(y - a)$. Find y when $a = 6$ and $b = 4$.

9–27. Solve $A = \dfrac{\theta \pi r^2}{360}$ for r. This is the formula for the area of a sector of a circle. Find r when $A = 20\pi$ and $\theta = 72$.

SOLVING COMPLEX FORMULAS

Many formulas are used in business and industry. However, research and technology brings new problems with new formulas to solve these problems. Thus, no matter how many formulas you may learn, either in school or on the job, there will always be new formulas. It is important for you to learn how to solve formulas in order to learn the procedures to solve future problems. Formulas become extremely complex, but once you learn the techniques of solving formulas, your confidence and ability will grow.

The next group of formulas are more complex. It is important to use the skills you have learned to solve these formulas and literal questions.

Example 9–15

Current Flowing Through Armature of a Generator

The amount of current flowing through the armature of a generator can be determined by the formula

$$I = \dfrac{E - e}{R}$$

PROBLEM: Solve this formula for E.

SOLUTION: Remove the fraction from the equation by multiplying each side of the equation by R. The numerator of the fraction represents more than one quantity. Remember to place parentheses around the quantity $(E - e)$.

$$I = \frac{(E - e)}{R}$$

$$(R)\,I = \frac{(E - e)}{\cancel{R}}\overset{1}{(\cancel{R})}$$

Add e to each side of the equation.

$$RI = E - e$$
$$e + RI = E - e + e$$
$$E = e + RI$$

Example 9–16

Drop in Voltage

The drop in voltage can be found by using the formula

$$\frac{E}{e} = \frac{R + r}{r}$$

PROBLEM: Solve for r.

SOLUTION: Clear the fraction from the equation by multiplying each side of the equation by the lowest common denominator, er. Remember to place parentheses around $(R + r)$.

$$\frac{E}{e} = \frac{R + r}{r}$$

$$(\cancel{e}r)\,\frac{E}{\cancel{e}} = \frac{(R + r)}{\cancel{r}}(e\cancel{r})$$

$$(r)\,E = (R + r)\,(e)$$
$$rE = Re + re$$

Subtract re from each side of the equation.

$$rE = Re + re$$
$$rE - re = Re + re - re$$
$$rE - re = Re$$

The quantities which contain r can be written in another way.

$$rE - re = Re$$
$$r(E - e) = Re$$

Now divide both sides by the quantity $(E - e)$ to find r.

$$\frac{r\cancel{(E - e)}}{\cancel{(E - e)}} = \frac{Re}{(E - e)}$$

$$r = \frac{Re}{E - e}$$

Example 9–17

Sum of a Geometric Progression

The sum of a geometric progression can be determined by the formula

$$S = \frac{r - a}{r - 1}$$

PROBLEM: Solve for r.

SOLUTION: Remove the fraction from the equation by multiplying each side by the lowest common denominator, $r - 1$.

$$S = \frac{r - a}{r - 1}$$

$$(r - 1)S = \frac{(r - a)}{\cancel{(r - 1)}}\cancel{(r - 1)}$$

$$(r - 1)S = r - a$$
$$rS - S = r - a$$

Thus add $+S$ and $-r$ to each side of the equation and solve.

$$rS - S + S - r = r - a + S - r$$
$$rS - S + S - r = r - a + S - r$$
$$rS - r = S - a$$

Rewrite $rS - r$ by factoring r from each term on the left side of the equation.

$$r(S - 1) = S - a$$

Now divide by the quantity $(S - 1)$ to find r.

$$r(S - 1) = S - a$$

$$\frac{r\cancel{(S - 1)}}{\cancel{(S - 1)}} = \frac{S - a}{(S - 1)}$$

$$r = \frac{S - a}{S - 1}$$

Chapter 9 / Formula Transposition

Example 9–18

> **Parallel Resistance**
> The parallel resistance can be found by the formula
> $$\frac{1}{R} = \frac{1}{r_1} + \frac{1}{r_2}$$

PROBLEM: Solve for r_1.

SOLUTION: Remove the fractions by multiplying each side of the equation by the lowest common denominator, Rr_1r_2.

$$\frac{1}{R} = \frac{1}{r_1} + \frac{1}{r_2}$$

$$Rr_1r_2\left(\frac{1}{R}\right) = (Rr_1r_2)\left(\frac{1}{r_1}\right) + Rr_1r_2\left(\frac{1}{r_2}\right)$$

$$r_1r_2 = Rr_2 + Rr_1$$

To get all of the terms that include r_1 on one side of the equation, subtract Rr_1 from each side of the equation.

$$r_1r_2 = Rr_2 + Rr_1$$
$$r_1r_2 - Rr_1 = Rr_2 + Rr_1 - Rr_1$$
$$r_1r_2 - Rr_1 = Rr_2$$

Then factor r_1 from each term on the left side of the equation.

$$r_1r_2 - Rr_1 = Rr_2$$
$$r_1(r_2 - R) = Rr_2$$

Divide each side of the equation by $(r_2 - R) = Rr_2$.

$$r_1(r_2 - R) = Rr_2$$

$$\frac{r_1(r_2 - R)}{(r_2 - R)} = \frac{Rr_2}{(r_2 - R)}$$

$$r_1 = \frac{Rr_2}{r_2 - R}$$

Study this example and carefully note each step. These procedures will help you solve other formulas.

EXERCISE 9–3

Solve the following problems. Review the procedures in the examples if you have any difficulty. Show all necessary work.

9–28. The formula for detemining latent heat vaporization is $Q = \frac{WL}{T}$. Solve the formula for T when $Q = 100$, W 5, and $L = 800$.

9–29. The formula for finding the average speed of a uniformly accelerating body is $V = \frac{V_t + V_o}{2}$. Solve the formulas for V_o and find V_o when $V = 312$ and $V_t = 112$.

9–30. The formula for determining temperature conversion is $T = \frac{1}{a} + t$. Solve the formula for a and find a when $T = 96$ and $t = 46$.

9–31. Using the electrical equivalent heat formula, $Q = 0.000477\ EIT$, solve for T and find T when $E = 1000$, $I = 200$, $Q = 95.4$.

9–32. Solve the thickness-of-pipe formula, $A = \dfrac{m}{t}(p + t)$, for t and find t when $A = 1$, $m = \dfrac{1}{4}$, and $p = 12$.

9–33. The formula for finding the area of an ellipse is $A = \pi ab$. Solve the formula for b and find b when $A = 113.04$ square units and $a = 18$.

9–34. Using the formula for finding the theoretical amount of air required to burn solid fuel, $M = 10.5C + 35.2\left(w - \dfrac{C}{8}\right)$, solve for C and find C when $M = 249.2$ and $w = 5$.

9–35. Solve the photographic enlargement formula, $\dfrac{1}{x} + \dfrac{1}{nx} = \dfrac{1}{f}$, for x and find x when $f = 12$ and $n = 3$.

9–36. The formula for finding the tap-size drill for a U.S. Standard thread is $S = T - \dfrac{1.299}{N}$. Solve the formula for N and find N when $T = 5.433$ and $S = 5$.

9–37. Solve the differential pulley formula, $W = \dfrac{2PR}{R - r}$, for R and find R when $P = 12$, $r = 75$, and $W = 48$.

9–38. Solve the algebraic equation, $x - y = xy$, for y and find y when $x = \dfrac{1}{8}$.

9–39 Using the prismoidal formula $V = \dfrac{h}{6}(B + 4M + b)$, solve for M and find M if $B = 25$, $b = 5$, $h = 36$, and $V = 420$.

9–40. Solve the formula for expansion of gases, $V_1 = V_0(1 + 0.5t)$, for t and find t when $V_1 = 112.5$ and $V_0 = 3$.

9–41. A formula for finding magnetic intensity is $H = \dfrac{0.4\pi NI}{L}$. Solve the formula for I and find I when $N = 4$, $L = 1.256$, $H = 2$, and $\pi = 3.14$.

Chapter 9 / Formula Transposition

9–42. Using the formulas for finding the perimeter of an ellipse, $P = \pi\sqrt{2(a^2 + b^2)}$, solve for b and find b when $P = 16\pi$ and $a = 9.59$.

9–43. Solve the formula for finding the width of a segment of a circle, $W = 2\sqrt{h(2r - h)}$, for r and find r when $h = 8$ and $W = 8\sqrt{8}$.

9–44. The formula for finding the volume of a torus is $V = 2\pi^2 Rr^2$. Solve the formula for r and find r when $R = 49$, $\pi = \dfrac{22}{7}$, and $V = 968$ square units.

9–45. The formula for finding the total surface area of a right circular cylinder is $T = 2\pi r(r + H)$. Solve for H and find H when $T = 320\pi$ and $r = 8$.

9–46. Solve the formula for finding the volume of a sphere, $V = \dfrac{\pi d^3}{6}$, for d and find d when $V = 288\pi$.

9–47. The formula for finding the distance an object travels when acted upon by gravity is $S = 16t^2 + vt$. Solve the formula for v and find v when $S = 1000$ feet and $t = 3$ seconds.

9–48. Using the formula for finding the distance an object travels when acted upon by gravity and an initial velocity, $S = 16t^2 + Vt$, solve for V and find V when $S = 1100$ and $t = 4$.

9–49. The formula for finding the exposed surface area of a cylinder is

$$S = \left(\dfrac{\pi d^2}{2} + \dfrac{\pi dl}{r}\right)\left(\dfrac{4rc}{d^2 l}\right)$$

Find S when $d = 4$, $l = 12$, $c = 2$, and $r = 2$. Leave the result in terms of π.

9–50. The formula for finding the volume of a cylinder is $V = \pi h(R + r)(R - r)$. Solve for V when $R = 8$, $h = 20$, and $r = 4$. Leave the answer in terms of π.

THINK TIME

Make a list of five formulas used in your occupation. Study the formulas carefully to assure your ability to work with them efficiently. Think of problems you may have on the job and how you may use the formulas.

Use the formulas to solve these problems. Ask your instructor or a friend to check your work.

Formula transposition skills are vital to many occupations; the following practice problems are for your further development.

EXERCISE 9–4

Solve the following problems for the given unknowns. Show all necessary work. Refer to the examples for help.

9–51. Solve $F = \dfrac{Nmv^2}{3}$ for v.

9–52. Solve $T = ph + 2A$ for h.

9–53. Solve $S = \dfrac{E - IR}{0.220}$ for I.

9–54. Solve $\dfrac{1}{f} = \dfrac{1}{p} + \dfrac{1}{q}$ for q.

9–55. Solve $I = \dfrac{En}{R + nr}$ for r.

9–56. Solve $E = RI + \dfrac{rI}{n}$ for I.

9–57. Solve $A = \dfrac{2}{3}hw$ for h.

9–58. Solve $V = 2\pi^2 R r^2$ for R.

9–59. Solve $V = \pi R^2 h - \pi r^2 h$ for r^2.

9–60. Solve $T = \dfrac{(P + p)s}{2}$ for p.

9-61. Solve $V = \dfrac{4\pi r^3}{3}$ for r.

9-62. From problem 9-61, find r when $V = 1333.333\pi$.

9-63. Given $F = \dfrac{16mx}{T^2}$, find x.

9-64. From problem 9-63, find x when $m = 6$, $T = 4$, and $F = 90$.

9-65. Solve $\dfrac{1}{R} = \dfrac{1}{r_1} + \dfrac{1}{r_2} + \dfrac{1}{r_3}$ for r_3.

9-66. Solve $w = \dfrac{2PR}{R-r}$ for R.

9-67. Solve $P = \pi\sqrt{2(a^2 + b^2)}$ for a.

9-68. From problem 9-67, find b when $P = 12\pi$ and $a = 4.796$.

9-69. Solve $S = 16t^2 + Vt$ for V.

9-70. Solve $V = \dfrac{\pi d^3}{6}$ for d.

PROCEDURES TO REMEMBER

1. To solve formulas:
   (a) Read the formula to understand the purpose of the formula.
   (b) Reread the formula, noting the unknown quantity that is to be found.
   (c) Determine what procedures you need to use to solve the formula for the unknown quantity.
   (d) Remember the "*scales of justice*" as you select the mathematical procedure to solve the formula. When you add, subtract, multiply, or divide or take a root on one side of the equation, you must perform the same operation on the other side of the equation.
   (e) Show all necessary steps and work as you perform the mathematical operations. This will help you to prevent careless errors.

(f) Check through your work to make certain that you have reached a reasonable solution and that your answer is stated in the correct form (i.e., square feet, cubic inches, etc.).

CHAPTER SUMMARY

Formula transposition is the process of solving a formula for the desired quantity by changing its original form mathematically.

CHAPTER TEST

T–9–1. Solve $P = 2l + 2w$ for w.

T–9–2. Solve $A = \dfrac{bh}{2}$ for b.

T–9–3. Solve $C = 2\pi r$ for r.

T–9–4. Solve $Q = 0.7\ ELT$ for L.

T–9–5. Solve $a - b = ab$ for b.

T–9–6. Solve $P = 2a + b$ for a.

T–9–7. Find s, to the nearest hundredth, when $S = 4000$ and $r = 462$ in the formula $S = \pi rs$ ($\pi = 3.14$).

T–9–8. Find r when $i = \$800$, $p = \$5000$, and $t = 2$ in the formula $i = prt$.

T–9–9. Solve $T = 2B + pH$ for H.

T–9–10. Solve $S = \dfrac{\pi rs}{2}$ for r.

T-9-11. Solve $C = \dfrac{5(F-32)}{9}$ for F.

T-9-12. Given the formula in problem T-9-11, find F if $C = 100$.

T-9-13. Solve $L = \dfrac{\pi r A}{180}$ for A.

T-9-14. Given the formula in problem T-9-13, find A when $L = 314$ and $r = 60$.

T-9-15. Solve $A = 6e^2$ for e.

T-9-16. Solve $A = \pi r^2$ for r.

T-9-17. Solve $S = 4\pi r^2$ for r.

T-9-18. Given the formula in problem T-9-17, find r when $S = 900\pi$.

T-9-19. Find y when $a = 6$ and $b = 4$ in the formula $(y - a) - b^2 = 2ab$. (First, solve the formula for y.)

T-9-20. Find e when $V = 343$ cubic inches given the formula $V = e^3$.

T-9-21. Solve $V = \dfrac{\pi r^2 h}{3}$ for h.

T-9-22. Solve $M = \dfrac{YI}{r}$ for r.

T-9-23. Solve $\dfrac{p_1 V_1}{T_1} = \dfrac{p_2 V_2}{T_2}$ for T_1.

T-9-24. Solve $A = \dfrac{h(B + b)}{2}$ for b.

T-9-25. Solve $F = \dfrac{4^2 n^2 W x}{g}$ for n.

10

RATIO AND PROPORTION

OBJECTIVES

1. To define ratio and proportion.
2. To develop skills solving ratios and proportions.
3. To solve applied ratios and proportions problems of technology, industry, and everyday life.

SELF-TEST

The completion of this self-test will indicate whether you need to further develop the skills taught in this chapter. Read each problem carefully and solve. Be sure to show your work so that errors may be identified. Write the ratios in fractional form and simplify.

S–10–1. 39 to 3 $\quad \frac{13}{1}$

S–10–2. $18y$ to $3y$ $\quad \frac{6}{1}$

S–10–3. $1\frac{3}{4}$ to $1\frac{1}{2}$ $\quad \frac{7}{4} \div \frac{3}{2} = \frac{7}{4} \times \frac{2}{3} = \frac{7}{6}$

S–10–4. 5 hours to 50 minutes $\quad \frac{6}{1}$

Solve the following proportions problems for the unknown.

S–10–5. $\frac{4}{10} = \frac{N}{40}$

$10N = 160$
$N = 16$

S–10–6. $\frac{4}{12} = \frac{Y}{9}$

$12y = 36$
$y = 3$

S–10–7. $\frac{2.5}{K} = \frac{10}{6}$

$10K = 15$
$K = 1.5$

S–10–8. $4:X = X:16$

$\frac{4}{X} = \frac{X}{16}$
$X^2 = 64$
$X = 8$

S–10–9. $27:x = 18:12$

$18x = 324$
$x = 18$

S–10–10. $y:15 = 75:15$

$15y = 1125$
$y = 75$

S–10–11. What number is to 14 as 2.5 is to 6?

$6x = 35$
$x = 5.8333...$

S–10–12. Nylon tricot sells at the rate of $8.37 for 3 yards. How much will $\frac{1}{2}$ yard cost?

$\frac{3}{8.37} = \frac{1/2}{x} = 3x = 4.185$
$x = \$1.40$

S–10–13. How much will 3 pounds of coffee cost if 2 pounds of coffee cost $3.28?

$\frac{3}{x} = \frac{2}{3.28} = 2x = 9.84$
$x = \$4.92$

S–10–14. A company needs five workers to produce 31 air conditioners per day. How many workers would be needed to produce 217 per day?

$\frac{5}{31} \; \frac{x}{217} = 31x = 1085$
$x = 35$

S–10–15. A motorist averages 450 miles per day at a speed of 45 miles per hour. If she increases her speed to 55 miles per hour, how many miles would she average per day?

550 miles

S–10–16. A stock pays a dividend of $5.50 per share every 182.5 days. What would the dividends amount to in 3 years?

$\frac{5.50}{182.5} = \frac{x}{1095} = 182.5x = 6022.5$
$x = \$33.00$

S–10–17. A restaurant serves 175 people on the average each evening. Of those served, 85 ordered steak. The restaurant owner remodeled so that he now serves 245 people on the average each evening. If the ratio remains the same, how many steaks will he serve?

$\frac{175}{85} \; \frac{245}{x} \; 175x = 20825$
$x = 119$

S–10–18. If you travel 594 miles in 11 hours, how long will it take to travel 486 miles?

$\frac{594}{11} \; \frac{486}{x} \; 594x = 5346$
$x = 9$ hours

S–10–19. A fast-food restaurant has an average daily gross sales of $4,000 and a daily labor cost of $450. If the labor cost increases to $486, what would the gross sales need to be to keep the same profit ratio?

$\frac{4000}{450} = \frac{x}{486} = 450x = 1944000$
$x = \$4320.00$

S–10–20. How far is it across the pond shown in Figure 10–1 from point A to point B?

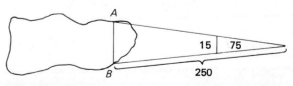

Figure 10–1

$\frac{15}{75} = \frac{x}{250} \quad 75x = 3750$
$x = 50$

Chapter 10 / Ratio and Proportion

S–10–21. A blueprint lists the scale as $\frac{1}{4}$ inch to 6 inches. Determine the height of a doorway that measures 4 inches on the blueprint.

$$\frac{\frac{1}{4}}{6} = \frac{4}{x} \quad \tfrac{1}{4}x = 24$$
$$x = 96 \text{ inches}$$

S–10–22. A copper wire is 360 feet long and has a resistance of 3.82 ohms. What is the resistance of 1000 feet of this wire?

$$\frac{360}{3.82} = \frac{1000}{x} \quad 360x = 3820$$
$$x = 10.61\ \Omega$$

S–10–23. The larger of two gears in mesh makes 35 revolutions per minute. The smaller gear has a diameter of 38 centimeters and makes 180 revolutions per minute. What is the diameter of the larger gear?

$$35x = 6840$$
$$x = 195.42\ \text{cm}$$

S–10–24. Shag shoots a score of 195 on a rifle range, using 25 rounds. What would his score be if he fired 35 rounds?

$$\frac{25}{195} = \frac{35}{x} \quad 25x = 6825$$
$$x = 273$$

S–10–25. Two speedboats start at the same time from the same place and travel in opposite directions. The ratio of their rates is 1:1.5. In 3 hours they are 60 miles apart. What is the rate of speed of the faster boat?

INTRODUCTION

We often compare one thing to another. For example, we may compare the gasoline consumption of a small car to that of a large car. If a small car uses 1 gallon of gasoline to travel 30 miles and a large car uses 1 gallon of gasoline to travel 15 miles, we could say that the small car travels twice as far as the large car on 1 gallon of gasoline. The ratio of miles per gallon is 2 to 1. We could also say that the ratio of miles traveled is 30 to 1 for the small car and 15 to 1 for the large car.

We use ratios to compare energy efficiency, gears, cost-effectiveness, inputs to outputs, and many other things in technology and industry and everyday situations.

RATIOS

A *ratio* is a comparison of one quantity to another. Similar comparisons are made when quantities are stated in fractions. Ratios always compare like quantities. As the saying goes, apples must be compared to apples and oranges to oranges—not apples to oranges.

Examples of ratios and how they may be stated are:

$$\frac{15}{60} = 15:60 \quad \text{or} \quad 15 \text{ is to } 60 = \frac{1}{4}$$

$$\frac{144}{12} = 144:12 \quad \text{or} \quad 144 \text{ is to } 12 = \frac{12}{1}$$

$$\frac{3 \text{ hours}}{45 \text{ minutes}} = \frac{\frac{3 \text{ hours}}{1}\left(\frac{60 \text{ minutes}}{1 \text{ hour}}\right)}{45 \text{ minutes}}$$

$$= \frac{180 \text{ minutes}}{45 \text{ minutes}} = \frac{4}{1}$$

When two quantities are compared, they must be in the same units. Ratios may be simplified by dividing equal quantities into each quantity. When a ratio is inverted, an *inverse ratio* results.

When ratios are simplified, they must be in the same units. If they are not in the same units, convert them to like units. Generally, this will require conversion to smaller units. Reducing should be only to whole numbers, not to decimals or fractions.

Example 10–1

   PROBLEM: What is the ratio of 12 to 48?

   SOLUTION: The ratio of 12 to 48 should be written in a fractional form and then reduced.

$$12:48 = \frac{12}{48} = \frac{1}{4}$$

Example 10–2

   PROBLEM: Write the ratio of 50 cents to $3.

   SOLUTION: Convert $3 to cents by multiplying $3 by 100, because there are 100 cents in 1 dollar. Then reduce to lowest terms.

$$50 \text{ cents} : 300 \text{ cents} = \frac{50}{300} = \frac{1}{6}$$

Example 10–3

   PROBLEM: What is the ratio of $5\frac{1}{2}$ to $\frac{1}{2}$?

   SOLUTION: Write the ratio as a fraction.

$$5\frac{1}{2} \text{ to } \frac{1}{2} = \frac{5\frac{1}{2}}{\frac{1}{2}} = 5\frac{1}{2} \div \frac{1}{2}$$

Complete the indicated division.

$$5\frac{1}{2} \div \frac{1}{2} = \frac{11}{2} \div \frac{1}{2} = \frac{11}{2} \times \frac{2}{1} = \frac{11}{1}$$

Example 10–4

   PROBLEM: What is the ratio of 6 hours to 45 minutes?

   SOLUTION: Convert 6 hours to minutes.

$$\left(\frac{6 \text{ hours}}{1}\right) \frac{60 \text{ minutes}}{1 \text{ hour}} = 360 \text{ minutes}$$

Set up the ratio and reduce to lowest terms.

$$\frac{360}{45} = \frac{360}{45} = \frac{40}{5} = \frac{8}{1}$$

EXERCISE 10–1

Simplify the following ratios by reducing to lowest terms.

10–1. $\frac{4}{8}$ *1/2*

10–2. $\frac{8}{24}$ *1/3*

10–3. $\frac{32}{24}$ *4/3*

10–4. $\frac{48}{12}$ *4/1*

10–5. $\frac{2.4}{0.3}$ *8/1*

10–6. $\frac{6\frac{1}{2}}{\frac{1}{2}}$ *13/1*

10–7. 7 to 42   *1/6*

10–8. 96 to 12   *8/1*

10–9. $\frac{5}{12}$ to $\frac{8}{15}$ *25/32*

10–10. 6.5 to 0.5 $\frac{13}{1}$

10–11. $\frac{3 \text{ minutes}}{1 \text{ hour}}$ $\frac{1}{20}$

10–12. 8 centimeters to 40 centimeters

$\frac{1}{5}$

10–13. 18 games to 2 games

9

10–14. 5 nickels to 4 dimes

$\frac{5 \times 5¢}{4 \times 10¢} = \frac{25}{40}$

$= \frac{5}{8}$

10–15. 6 feet to 144 inches   6 ft to 12 ft

$\frac{1}{2}$

10–16. 5 days to 6 weeks

$\frac{5}{42} =$

10–17. $10 to 25 cents

$\frac{10 \times 100¢}{25¢} = \frac{40}{1}$

10–18. 300 turns to 37.5 turns

$\frac{8}{1}$

10–19. a^2b^2 to ab

$\frac{AB}{1}$

10–20. 6 feet 6 inches to 3 yards

13)216
  78
  108

$\frac{6.5}{9}$

$\frac{13}{18}$

PROPORTIONS

When one ratio is equal to another ratio, a proportion is formed. A *proportion* is a statement of equality between two ratios.

There are several ways that a proportion may be stated. Two of the more common are:

Proportional form: $2:3 = 8:12$

Fraction form: $\dfrac{2}{3} = \dfrac{8}{12}$

A proportion has four terms. Using the example above, 2 is the *first term*, 3 is the *second term*, 8 is the *third term*, and 12 is the *fourth term*. The product of the first and fourth terms is called the product of the *extremes* (outside terms), and the product of the second and third terms is called the product of the *means* (inside terms). The product of the means equals the product of the extremes.

There are times when only three of the four terms of a proportion are known. To determine the fourth term, find the product of the extremes and set it equal to the product of the means. This process is often referred to as setting the cross products equal to each other, or *cross multiplication*. Once the products have been found, solve for the unknown.

Example 10–5

PROBLEM: Solve the proportion $X:8 = 6:24$ for the unknown.

SOLUTION: Find the product of the means and the product of the extremes.

$$X:8 = 6:24 \qquad (X)(24) = (8)(6)$$

(means and extremes indicated)

Divide both sides by 24.

$$(X)(24) = (8)(6)$$
$$\dfrac{24X}{24} = \dfrac{(8)(6)}{24}$$
$$X = 2$$

Then check the results as shown:

$$2:8 = 6:24 \qquad \begin{array}{c}(2)(24) = (8)(6)\\ 48 = 48\end{array}$$

Example 10–6

PROBLEM: Solve the proportion for the unknown.

$$\dfrac{2}{3} = \dfrac{K}{12}$$

SOLUTION: Find the cross products of the fractional proportion. This operation is called *cross multiplying*.

$$\dfrac{2}{3} = \dfrac{K}{12}$$

$$(2)(12) = (3)(K)$$

Solve for K by dividing both sides by 3.

$$\dfrac{\overset{4}{\cancel{(2)(12)}}}{\cancel{3}} = \dfrac{\overset{1}{\cancel{(3)}}(K)}{\cancel{3}}$$
$$8 = K$$

Check the value as follows:

$$\dfrac{2}{3} = \dfrac{8}{12}$$

$$(2)(12) = (8)(3)$$
$$24 = 24$$

Example 10–7

PROBLEM: Solve the proportion for the unknown.

$$\dfrac{30}{5} = \dfrac{10}{X}$$

SOLUTION: Find the cross products of the fractional proportion.

$$\dfrac{30}{5} = \dfrac{10}{X}$$
$$(30)(X) = (10)(5)$$

Solve for X by dividing both sides by 30.

$$\dfrac{\cancel{(30)}(X)}{\cancel{30}} = \dfrac{(10)(5)}{30}$$

$$X = \dfrac{5}{3} \text{ or } 1.\overline{6}$$

Chapter 10 / Ratio and Proportion

Check the results as follows:

$$(30)(X) = (10)(5)$$

$$\cancel{(30)}^{10}\left(\frac{5}{\cancel{3}_1}\right) = (10)(5)$$

$$50 = 50$$

EXERCISE 10-2

Solve and check the following proportions.

10–21. $\dfrac{5}{8} = \dfrac{j}{24}$

$8j = 124$
$j = 15$

10–22. $\dfrac{3}{7} = \dfrac{X}{49}$

$7x = 147$
$x = 21$

10–23. $\dfrac{80}{r} = \dfrac{60}{5}$

$60r = 400$
$R = 6.666...$

10–24. $\dfrac{60}{360} = \dfrac{k}{45}$

$360k = 2700$
$K = 7.5$

10–25. $\dfrac{10}{s} = \dfrac{0.4}{36}$

$.4s = 360$
$s = 900$

10–26. $\dfrac{y}{15} = \dfrac{40}{3}$

$3y = 600$
$y = 200$

10–27. $\dfrac{6}{x} = \dfrac{27}{36}$

$27x = 216$
$x = 8$

10–28. $\dfrac{0.3}{24} = \dfrac{0.2}{a}$

$.3A = 4.8$
$A = 16$

10–29. $4:2 = 13:x$

$4x = 26$
$x = 6.5$

10–30. $16:y = 64:8$

$64y = 128$
$y = 2$

10–31. $x:40 = \dfrac{1}{4}:2$

$2x = 10$
$x = 5$

10–32. $y:12 = 9:27$

$27y = 108$
$y = 4$

10–33. $112:128 = 42:x$

$112x = 5376$
$x = 48$

10–34. $36:9 = 28:x$

$36x = 252$
$x = 7$

10–35. $\dfrac{x-4}{20} = \dfrac{14}{70}$

$70x - 280 = 280$
$70x = 560$
$x = 8$

10-36. $\dfrac{9}{x} = \dfrac{x}{1}$

$x^2 = 9$

$x = 3$

10-37. $\dfrac{36}{4x} = \dfrac{9x}{49}$

$36x^2 = 1764$

$x^2 = 49$

$x = 7$

10-38. $\dfrac{x}{4.3} = \dfrac{19.7}{21.1775}$

$21.1775x = 84.71$

$x = 4$

10-39. $\dfrac{45}{5y} = \dfrac{9y}{64}$

$45y^2 = 2880$

$y^2 = 64$

$y = 8$

10-40. $\dfrac{0.400}{0.014} = \dfrac{0.08}{p}$

$.400\,p = .00112$

$p = .0028$

APPLICATIONS

Many occupations require the skill to solve problems involving ratios and proportions. The following examples relate to some of the applications in industry.

Example 10-8

PROBLEM: If a copying machine produces 48 copies in 30 seconds, how many seconds will it take to produce 240 copies?

SOLUTION: Set up the proportion in fractions and let X represent the time it will take to produce 240 copies.

Solve the proportion; remember that cross products are equal.

$$\dfrac{48}{30} = \dfrac{240}{X}$$

$(48)(X) = (30)(240)$

Divide each side of the equation by 48.

$$\dfrac{\cancel{(48)}(X)}{\cancel{48}} = \dfrac{\cancel{(30)(240)}}{\cancel{48}}$$

$X = 150$

It will take 150 seconds to produce 240 copies. Check your results as shown.

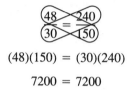

$(48)(150) = (30)(240)$

$7200 = 7200$

Example 10-9

PROBLEM: A man wishes to lift a rock ~~a rock~~ out of a shallow hole. He estimates the rock weighs 420 pounds. If a 12-foot plan (lever) is used, where will the fulcrum need to be placed if the man used 180 pounds?

SOLUTION: Recall the principle of the lever.

W_1 W_2

↓ d_1 d_2 ↓

▲

$(W_1)(d_1) = (d_2)(W_2)$

Then draw a sketch of the problem.

420 lb   180 ~~810~~ lb
(rock)   (man)
↓   ↓
d_1 ▲ $12 - d_1$

W_1 = weight to the left (420 lb)

d_1 = distance to the left (d_1)

W_2 = weight to the right (180 lb)

d_2 = distance to the right ($12 - d_1$)

Using the principle of the lever, write the problem as shown, where d_1 is the distance to the fulcrum from the rock and $12 - d_1$ is the distance of the man from the fulcrum.

$$(420)(d_1) = (12 - d_1)(180)$$

Then solve for d_1.

$$420d_1 = 12(180) - 180d_1$$
$$420d_1 + 180d_1 = (12)(180)$$
$$600d_1 = (12)(180)$$
$$\frac{\overset{1}{\cancel{600}}d_1}{\underset{1}{\cancel{600}}} = \frac{\cancel{(12)}\overset{\overset{18}{\cancel{180}}}{\cancel{(180)}}}{\underset{\underset{5}{\cancel{50}}}{\cancel{600}}}$$
$$d_1 = \frac{18}{5}$$
$$d_1 = 3.6$$

Thus the fulcrum should be 3.6 feet from the rock. Check as shown.

$$(420)(3.6) = (12 - 3.6)(180)$$
$$1512 = (8.4)(180)$$
$$1512 = 1512$$

Example 10–10

PROBLEM: A brick wall measures $3\frac{1}{2}$ inches on a drawing. How long is the wall if the scale of the drawing is $\frac{1}{4}$ inch = 25 feet?

SOLUTION: Let x = the actual length. Set up the proportion.

$$\frac{\text{Actual size}}{\text{Scale size}} = \frac{\text{actual length}}{\text{drawing length}}$$

$$\frac{25 \text{ ft}}{\frac{1}{4} \text{ in.}} = \frac{x}{3\frac{1}{2} \text{ in.}}$$

Note: Convert the fractions to decimals.

$$\frac{25 \text{ ft}}{0.25 \text{ in.}} = \frac{x}{3.5 \text{ in.}}$$

Solve for x.

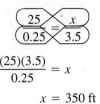

$$\frac{(25)(3.5)}{0.25} = x$$

$$x = 350 \text{ ft}$$

Check as shown.

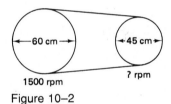

$$(25)(3.5) = (350)(0.25)$$
$$875 = 875$$

Example 10–11

PROBLEM: Two pulleys are connected by a belt. The diameter of the larger pulley is 60 centimeters and the diameter of the smaller pulley is 45 centimeters. If the larger pulley rotates at 1500 revolutions per minute (rpm), how fast does the smaller pulley rotate?

SOLUTION: Sketch the pulleys (see Figure 10–2) and set up a proportion. Remember that pulley and gears are to be set up as an inverse ratio.

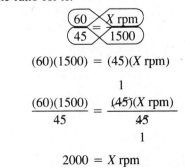

Figure 10–2

$$\frac{\text{Diameter of larger}}{\text{Diameter of smaller}} = \frac{\text{revolutions per minute of smaller}}{\text{revolutions per minute of larger}}$$

Let X be the revolutions per minute of the smaller pulley and set up the problem.

$$\frac{60}{45} = \frac{X \text{ rpm}}{1500}$$

Solve the ratio for X.

$$\frac{60}{45} \bowtie \frac{X \text{ rpm}}{1500}$$

$$(60)(1500) = (45)(X \text{ rpm})$$

$$\frac{(60)(1500)}{45} = \frac{\overset{1}{\cancel{(45)}}(X \text{ rpm})}{\underset{1}{\cancel{45}}}$$

$$2000 = X \text{ rpm}$$

Check as shown.

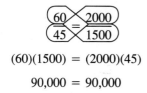

(60)(1500) = (2000)(45)

90,000 = 90,000

Note: Another way to state the pulley or gear principle is: The product of the diameter of the larger pulley times the rpm of the larger pulley equals the product of the diameter of the smaller pulley times the rpm of the smaller pulley. This is stated as $(D)(RPM) = (d)(rpm)$, where the capital letters indicate the larger pulley and the lowercase letters the smaller pulley.

EXERCISE 10–3

Work the following problems that are similar to applied problems in industry and daily living.

10–41. A baker sells 3 doughnuts for $0.79. How much will $1\frac{1}{2}$ dozen doughnuts cost?

10–42. A laborer is paid at a rate of $237 for every 3 days that he works. How much will he earn in 21 days?

10–43. A drawer is $1\frac{1}{2}$ feet wide. In a drawing the drawer is 1 inch wide. Determine the scale ratio of the actual width of the drawer to the scale width of the drawer in the drawing.

10–44. If 12 workers can complete a job in 75 days, how many workers would be required to complete the job in 10 days? (This is an application of an inverse ratio. See Example 10–11.)

10–45. The scale on a map is 1 inch = 12 miles. How many miles is it from one town to another if the distance measures $7\frac{1}{4}$ inches on the map?

10–46. In Arkansas, the ratio of harvested land to the total area of the state is 1:4.8. If the total area of the state is 53,225 square miles, how many square miles comprise harvested land?

10–47. A machine can produce 60,000 nails in 12 hours. At this rate, how long would it take to produce 160,000 nails?

10–48. The population of a certain country is 50,000,000. The area of the country is 215,000 square miles. Determine the population density (number of people per square mile).

10–49. A train travels 84 miles in $1\frac{2}{5}$ hours. How long will it take to travel 480 miles?

10–50. If 6 workers take $10\frac{1}{2}$ days to complete a job, how long will it take 10 workers? (*Remember:* Workers × time worked = workers × time worked.)

10–51. A cubic foot of water weighs 62.5 pounds and a cubic foot of ice weighs 57.5 pounds. What is the ratio of the weight of ice to water?

10–52. For Figure 10–3, find BC when $AD = 8$, $AB = 18$, and $DE = 6$.

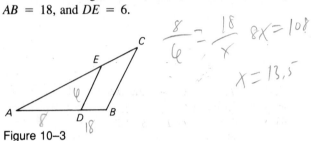

Figure 10–3

10–53. A picture is 24 inches long and 18 inches wide. If a reduced copy of this picture is 5.5 inches long, determine its width.

10–54. The volume of a given amount of gas is inversely proportional to the pressure on the gas. If a gas occupies 175 cubic inches when under a pressure of 20 pounds per square inch, what will be its volume if the pressure is decreased to 11 pounds per square inch?

10–55. The weight of dried apples to the weight of fresh apples is in the ratio of 2:5. How many pounds of fresh apples are needed to produce 490 pounds of dried apples?

10–56. A 40-acre field yields 880 bushels of wheat. When fertilizer is applied, the yield increased by $1\frac{1}{2}$ times. How much yield could be expected from an 84-acre field when fertilizer is applied?

10–57. A large gear with 30 teeth is turning at 400 revolutions per minute and turns a smaller gear with 12 teeth. If the revolutions per minute of the larger gear are tripled, find the speed of the smaller gear.

10–58. If 16 pounds of cement are used to make 80 pounds of concrete, how many pounds of concrete can be made with 120 pounds of cement?

10–59. A scientific principle of machinery concerning the stretching of springs according to weight is known as *Hooke's Law*. It is stated as a proportion as follows:

$$\frac{\text{First weight}}{\text{First distance stretched}} = \frac{\text{second weight}}{\text{second distance stretched}}$$

A spring is stretched 0.5 inch by a 15.5-pound weight. How far will it be stretched by a 90-pound weight?

$$\frac{15}{15.5} = \frac{x}{90} \quad 15.5x = 45$$
$$x = 2.9 \text{ IN}$$

10–60. A small pulley, 4.5 inches in diameter, is belted to a 16-inch-diameter pulley. The smaller pulley is turning at 3,500 revolutions per minute and is decreased by $\frac{1}{7}$; how fast is the larger pulley turning after the increase?

$$16 \times x \qquad 4.5 \times 3500$$
$$16x = 15750$$
$$x = 984 \times \frac{6}{7}$$
$$x = 843.75 \text{ RPM}$$

10–61. A pry bar uses the same principles as the lever. If a bar is 60 inches long and fulcrum is placed 6 inches from the end of a 1500-pound weight, how much weight must be applied to lift the weight?

[diagram: 60, 6in, 1500]

$$1500 \times 6 = (54) W$$
$$W = 166.6 \text{ LB}$$

10–62. Find the distance from the fulcrum to the base of a 2500-pound milling machine if the pry bar is 80 inches long and 220 pounds of force are applied.

$$2500x = 80 - x (220)$$
$$2500x = 17600 - 220x$$
$$17600 = 2720x \qquad x = 6.47 \text{ IN}$$

10–63. A claw hammer, used to pull a nail, uses the principle of the lever. (See Figure 10–4.) If 80 pounds of force (W_2) are applied on the handle which is 12 inches long, and if the distance from the claws to the nail is $2\frac{1}{4}$ inches, how much force can be applied to pull the nail?

10–64. Referring to Figure 10–4, how much force is needed if the nail's downward force is 224 pounds?

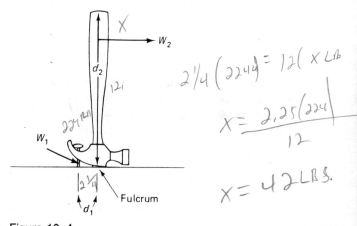

Figure 10–4

$$2\frac{1}{4}(224) = 12(x \text{ LB})$$
$$x = \frac{2.25(224)}{12}$$
$$x = 42 \text{ LBS.}$$

10–65. The breaking force of the hammer handle (problem 10–63) is 125 pounds. What is the maximum pulling force that could be obtained from the hammer?

10–66. A chef prepares a banquet for 60 people. He uses 18 pounds of potatoes. How many pounds of potatoes would he need for a banquet of 210 people?

$$\frac{60}{18} \quad \frac{210}{x} \qquad 60x = 3780$$
$$x = 63$$

Chapter 10 / Ratio and Proportion

10–67. A painter uses 1 quart of paint to cover 225 square feet of concrete wall. How many gallons would she need to cover 1800 square feet?

10–68. Thunder is heard 48 seconds after lightning flashes 10 miles away. When would a person 6 miles away hear the thunder?

10–69. What will it cost to serve 99 people ground beef that cost $1.39 per pound if 5 pounds will serve 20 people?

10–70. The maximum height reached by a bullet fired from a certain gun is proportional to the square of the time it takes to reach that height. If a bullet reaches a height of 5 feet 0.01 second after it is fired, what height would a bullet reach in half a second?

THINK TIME

Ratios and proportions may be used to solve a wide variety of problems. For example, they may be used to find screw thread depth, horsepower, mechanical efficiency of engines, taper and diameter, speed of pulleys and gears, strength of materials, velocity, density, and pressure. Think of ways in which you can use the principles of ratio and proportion to solve problems.

One of the common uses of ratio is to determine the portion of one's home that can used as a business expense for tax purposes. Follow the example given below.

PROBLEM: Medora owns her home and uses one room exclusively as an office for her service station business. In this room she keeps her accounts and makes out required reports; at night she uses it to receive calls for emergency road service. She determines that the size of the room is 120 square feet. The total square footage of her home is 1440. If her electric, fuel, and water bills total $3672 for 1 year, how much of this can be allowed for the office as business expenses?

SOLUTION: First, set up a proportion.

$$\frac{120 \text{ sq ft}}{1440 \text{ sq ft}} = \frac{X}{\$3672}$$

Find the product of the means and the product of the extremes and divide to determine the amount allowed.

$$(12)(\$3672) = (X)(1440)$$

$$\frac{(120)(\$3672)}{1440} = X$$

$$X = \$306$$

Check as shown.

$$\frac{120}{1440} = \frac{\$306}{\$3672}$$

$$\$440,640 = \$440,640$$

PROCEDURES TO REMEMBER

1. To simplify a ratio:
   (a) Convert to like terms.
   (b) Write as a fraction.
   (c) Reduce to lowest terms.
2. To solve a proportion:
   (a) Set up the proportion, converting to like terms if necessary.
   (b) Determine the product of the means and the product of the extremes.
   (c) Solve for the unknown.
   (d) Check by substituting the result for the unknown and cross multiply.

CHAPTER SUMMARY

1. A *ratio* is a comparison of one quantity to another.
2. Ratios always compare like items.
3. When two quantities are compared, they must be like units.
4. A *true ratio* is found by reducing to the lowest term.
5. When a ratio is inverted, an *inverse ratio* results.
6. Simplification of ratios should be to whole numbers, not to decimals or fractions.
7. A *proportion* is a statement of equality between two ratios.

8. Two common ways of expressing a proportion are:
   (a) Proportional form (3:4 = 9:12)
   (b) Fractional form $\left(\dfrac{3}{4} = \dfrac{9}{12}\right)$
9. A proportion has four terms. In the example 6:8 = 18:24, 6 is the *first term*, 8 is the *second term*, 18 is the *third term*, and 24 is the *fourth term*.
10. The product of the first and fourth terms is called the product of the *extremes* (outside terms).
11. The product of the second and third terms is called the product of the *means* (inside terms).
12. The product of the means equals the product of the extremes.

CHAPTER TEST

Write in fractional form and simplify problems.

T–10–1. 4 inches to 12 inches

$\dfrac{4}{12} \quad \dfrac{1}{3}$

T–10–2. 7 to 63

$\dfrac{7}{63} \quad \dfrac{1}{9}$

T–10–3. 5 feet 8 inches to 6 feet 2 inches

$\dfrac{5\tfrac{8}{12}}{6\tfrac{1}{6}} \quad \dfrac{68}{74} \quad \dfrac{34}{37}$

T–10–4. $\dfrac{6\tfrac{1}{2}}{7\tfrac{7}{8}}$

$\dfrac{13}{2} \times \dfrac{8}{7} = \dfrac{52}{7}$

T–10–5. $\dfrac{x}{7} = \dfrac{1.5}{1}$

$x = 10.5$

T–10–6. $\dfrac{1.5}{2} = \dfrac{3.5}{p}$

$1.5p = 7$
$p = 4.666...$

T–10–7. $\dfrac{E}{10.5} = \dfrac{9}{31.5}$

$31.5 E = 94.5$
$x = 3$

T–10–8. $\dfrac{84}{20} = \dfrac{21}{N}$

$84 N = 420$
$x = 5$

T–10–9. $j:8 = 18:j$

$j^2 = 144$
$j = 12$

T–10–10. $18:21 = M:63$

$21m = 1134$
$m = 54$

T-10-11. What number is to 28 as 5 is to 12?

$\frac{x}{28} = \frac{5}{12}$ $12x = 140$
   $x = 11.666...$

T-10-12. A trucker finds that he uses 63.5 gallons of fuel to drive 762 miles. How many gallons will he need to drive 4575 miles?

$\frac{63.5}{762} = \frac{x}{4575}$ $762x = 290,512.5$
   $x = 381.25$

T-10-13. Burlap material sells at a rate of $5.88 for 3 yards. How much will $6\frac{1}{2}$ yards cost?

$\frac{5.88}{3} = \frac{x}{\frac{13}{2}}$ $3x = 38.22$
   $x = \$12.74$

T-10-14. Dividends of $0.375 per share are paid every 90 days. How many days would it take to accumulate a total of $375.00 on 200 shares?

$\frac{.375}{1} = \frac{1.875}{x}$
$\frac{D}{375} = \frac{90}{.375 \times 200}$ $D = \frac{33750}{75}$
   $D = 450$

T-10-15. What is the ratio of two gears if one gear has 105 teeth and the other has 15 teeth?

$\frac{105}{15}$ $\frac{7}{1}$

T-10-16. Joel shoots a score of 164 on a rifle range using 30 rounds. What would his score be if he fired 45 rounds?

$\frac{164}{30} = \frac{x}{45}$ $30x = 7380$
   $x = 246$

T-10-17. A jet plane travels 3972 miles in 6 hours. At this rate, how far can the plane travel in $4\frac{1}{2}$ hours?

$\frac{3972}{6} = \frac{x}{4.5}$ $6x = 17874$
   $x = 2979$ miles

T-10-18. A florist plants 475 tulip bulbs, of which 456 survive for sale. The following year the florist wishes to have 600 tulips for sale. How many tulips will need to be planted at the same survival rate?

$\frac{475}{456} = \frac{x}{600}$ $456x = 285000$
   $x = 625$

T-10-19. If there are 160 pounds of nickel in 1 ton of alloy, how much nickel would be required to make 4500 pounds of the alloy?

$\frac{160}{2000} = \frac{x}{4500}$ $2000x = 720000$
   $x = 360$ pds

T-10-20. A copying machine produces 35 copies in 20 seconds. How many copies will it produce in 5 minutes?

$\frac{35}{20} = \frac{x}{300}$ $20x = 10500$
   $x = 525$

T–10–21. Find the missing dimension of the A-frame cottage shown in Figure 10–5.

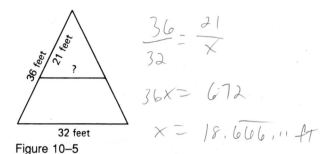
Figure 10–5

$$\frac{36}{32} = \frac{21}{x}$$

$36x = 672$

$x = 18.666\ldots$ ft

T–10–22. The scale on a drawing is $\frac{1}{8}$ inch = 3 feet. Determine the length in feet of a rafter if it measures $1\frac{3}{8}$ inches on the drawing.

$$\frac{\frac{1}{8}}{3} = \frac{\frac{11}{8}}{x}$$

$1/8 \, x = 49.5$

$x = \frac{396}{12}$

$x = 33$ ft

T–10–23. A 5-inch pulley makes 1800 revolutions per minute and drives a larger pulley at 300 revolutions per minute. What is the diameter of the larger pulley?

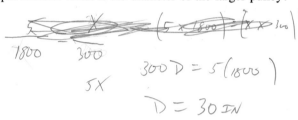

$300 D = 5(1800)$

$D = 30$ IN

T–10–24. A photograph has a length of 16 inches and width of 11 inches. An enlargement is made having a length of 45 inches. Determine the width of the enlargement.

$$\frac{16}{11} = \frac{45}{x}$$

$16x = 495$

$x = 30.94$ IN.

or 31 IN.

T–10–25. A chef uses 22 pounds of flour to bake the cakes he serves during the day. He generally serves cake to about 160 people. He decides to cut the price he charges for a piece of cake. Now 276 people order cake each day. How much flour does he use?

$$\frac{22}{160} \quad \frac{x}{276}$$

$160x = 6072$

$x = 37.95$ pounds

section three: applied geometry 11

BASIC DEFINITIONS AND PROPERTIES OF GEOMETRY

OBJECTIVES

1. To review the fundamental foundations of geometry.

2. To develop the skills required to solve basic applied geometry problems.

SELF-TEST

This test will measure your understanding and ability to work with basic symbols and definitions of geometry. If you complete the questions and problems successfully, you are ready to move on to more difficult geometric concepts. It is very important that you have a firm grasp of these geometric symbols and definitions before you start applied geometry problems.

S–11–1. The symbol $\perp$ indicates _____.

S–11–2. The symbol $\overleftrightarrow{AJ}$ means _____.

S–11–3. Write the symbol for line *NL*.

S–11–4. Write the symbol for angle *AOK*.

S–11–5. How long is a line?

S–11–6. If you were sailing due north and made a 270° turn to the right, in what direction would you be going?

S-11-7. How many degrees would the needle of the compass rotate if you turned from north to south?

S-11-8. Three fourths of a complete revolution gives an angle of _____ degrees.

S-11-9. Minutes used to measure angles are subdivided into _____.

S-11-10. The symbol for a straight line is _____.

S-11-11. A triangle with all sides of equal length is called an _____ triangle.

S-11-12. What angle do the hands of a clock make at 7 o'clock?

S-11-13. Measure with a protractor angle *NOP* shown in Figure 11-1.

S-11-14. Measure with a protractor angle *J* shown in Figure 11-2.

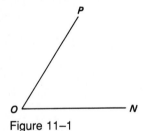

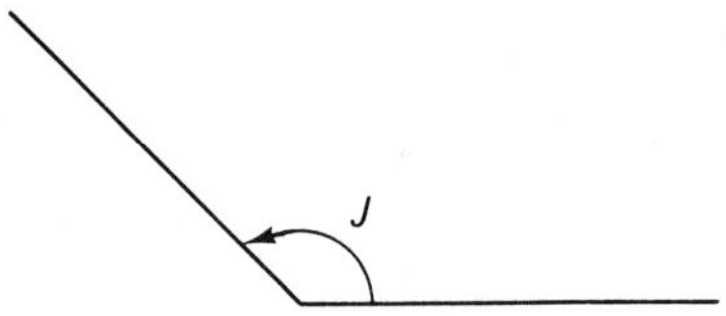

Figure 11-1

Figure 11-2

S-11-15. Construct a 73° angle with a protractor.

S-11-16. Construct a 153° angle with a protractor.

S-11-17. Find the complement of a 36° angle.

S-11-18. Find the supplement of a 67° angle.

S–11–19. How many degrees are there in one-fourth of a right angle?

S–11–20. Measure ∠WOZ in Figure 11–3.

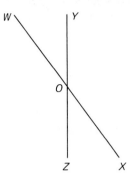

Figure 11–3

S–11–21. Measure ∠WOY in Figure 11–3.

S–11–22. Angles WOZ and ZOX are called _____ angles.

S–11–23. Measure ∠PON in Figure 11–4.

S–11–24. Measure ∠POM in Figure 11–4.

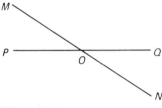

Figure 11–4

S–11–25. Measure ∠POQ in Figure 11–4.

INTRODUCTION

Geometry is an ancient branch of mathematics. Its oldest roots can be traced back five or six thousand years to the Egyptian and Babylonian civilizations. The precise construction of the pyramids is a monument to the use of geometry and skill of the Egyptians. These skills were also used to reestablish boundary lines that washed away during the flooding of the Nile river.

The first known geometer was a Greek mathematician, Euclid, who lived about 300 B.C. Euclid gathered the geometric principles, combined them with his own concepts, and wrote the first geometry book. This book included geometric principles that have changed very little over the centuries. In fact, Euclid's book could be used today with few changes. It is a rarity that any book of science or mathematics could be useful for such a long period of time. But Euclid wrote a nearly perfect geometry book even though he worked with the simplest of tools and equipment to make his calculations. His knowledge and skills of geometry make our work and life easier.

DEFINITIONS

The three basic properties of geometry are point, line, and plane. A *point* is a position or location that has no size or dimension. Generally, a point is represented by a dot with a capital letter beside it (.P). A *line* may be defined as a series or set of points extending indefinitely in opposite directions and may form a straight, broken, or curved line. A line does not have width or depth. It has length only and is generally identified by a lowercase letter beside it:

$\underset{a}{\text{———}}$ straight line; $\underset{b}{\wedge}$ broken line; $\underset{c}{\frown}$ curved line.

A *straight line* is the shortest distance between two points. Generally, when the word "line" is used, it means a straight line. The symbol for a straight line is $\overset{\leftrightarrow}{b}$. This symbol indicates that straight line b extends indefinitely in both directions. It is understood that a line extends indefinitely; thus the arrows are not added. Two lines that never meet, no matter how far they are extended, are called *parallel lines* and are identified by the symbol ∥. A *plane* is a flat surface that has two dimensions, length and width. The sides of a box are plane surfaces.

There are two types of geometry: plane and solid. *Plane geometry* deals with two dimensions: length and width. *Solid geometry* includes three dimensions: length, width, and height. Plane geometry is like two-dimensional art such as painting, whereas solid geometry may be likened to a three-dimensional art sculpture. The basic definitions given above will be of value when you consider the applied geometric concepts and forms.

PLANE FIGURES

As stated previously, a plane is a flat surface that extends indefinitely in all directions. A *plane figure* is a figure drawn on a flat surface. A *plane closed figure* is a figure that encloses an area or surface. Any plane figure can be described by line segments, points, and angles. Observe the plane closed figures illustrated in Figure 11–5. Note that they may be drawn with straight or curved lines.

Figure 11–5

There are names for plane figures. A *polygon* is a plane figure formed by straight-line segments. A *triangle* is any polygon formed by three straight-line segments. A *right triangle* is a triangle containing a right angle. An *equilateral triangle* is a triangle with all three sides of equal length. An *isosceles triangle* is a triangle with two equal sides. Observe the illustration of the triangles in Figure 11–6.

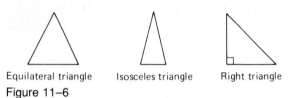

Equilateral triangle Isosceles triangle Right triangle
Figure 11–6

Four-sided polygons are formed with parallel lines and have special names. A *parallelogram* is a four-sided plane figure whose opposite sides are parallel. A *rectangle* is a parallelogram containing all right angles. A *rhombus* is a parallelogram with all sides of equal length. A *square* is a parallelogram with all equal sides and all right angles. Observe the illustrations of parallelograms in Figure 11–7.

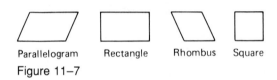

Parallelogram Rectangle Rhombus Square
Figure 11–7

SOLID FIGURES

Observe the drawing of a block shown in Figure 11–8. Because it has length, width, and height, it is called a *rectangular solid*. A solid is a figure that has depth, length, and width. Letters indicate the corners of the block.

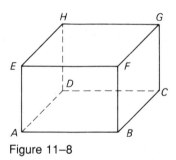

Figure 11–8

The line from point A to point B is called a *line segment* because it has *endpoints* A and B. The symbol for line segment AB is $\overline{AB}$. This means that line segment AB is only a portion of a line that extends indefinitely in both directions. Line segment AB is referred to as the edge of a rectangular solid.

Section Three / Applied Geometry

Lines $\overline{AB}$, $\overline{BF}$, $\overline{FE}$, and $\overline{EA}$ form a plane or a *face* of the rectangular solid. The plane is called *ABFE*. Point *F*, where the lines *EF*, *BF*, and *GF* meet and where planes *ABFE*, *BCGF*, and *EFGH* meet, is called a *vertex* of the rectangular solid.

Example 11-1

PROBLEM: Sketch a rectangular solid, letter it, and determine its planes, line segments, and vertices.

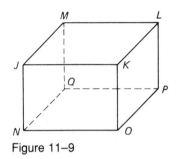

Figure 11-9

SOLUTION: First, draw the rectangular solid. Second, letter the rectangular solid. See Figure 11-9. Third, determine the planes. The planes are *JKLM*, *JKON*, *JMQN*, *KLPO*, *LMQP*, and *NOPQ*. Fourth, determine the line segments. The line segments are *JK*, *JN*, *JM*, *KL*, *KO*, *LM*, *LP*, *MQ*, *NO*, *NQ*, *OP*, and *PQ*. Fifth, indicate the eight vertices. The vertices are

J, formed by planes *JKLM*, *JKON*, *JMQN*
K, formed by planes *JKLM*, *JKON*, *KIPO*
L, formed by planes *JKLM*, *LMQP*, *KLPO*
M, formed by planes *JKNO*, *JMQN*, *LMQP*
N, formed by planes *JKNO*, *JMQN*, *LMQP*
O, formed by planes *JKNO*, *KLPO*, *NOPQ*
P, formed by planes *LMQP*, *KLPO*, *NOPQ*
Q, formed by planes *JMNQ*, *LMQP*, *NOPQ*

EXERCISE 11-1

Complete the following exercises to give you an opportunity to learn the basic geometric definitions.

11-1. List three types of parallelograms.

11-2. List three types of triangles.

11-3. Draw and label a line segment.

11-4. Draw and label a rectangular solid.

11-5. What are the three dimensions of a solid geometry?

11-6. A triangle with all three sides of equal length is called a _____.

11-7. Draw the symbol for parallel lines.

11-8. Draw the symbol for straight line *f*.

11-9. A plane closed figure formed by straight line segments is called a _____.

11-10. A triangle containing a right angle is called _____.

ANGLES

When two lines intersect (cross) each other, angles are formed. The point of intersection is called the *vertex* of the angles and is indicated by a capital letter. See Figure 11-10. Lines *AB* and *CD* intersect at point *O*. Point *O*

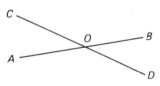

Figure 11-10

is the vertex of the angles. Each angle is indicated by the angle symbol (∠) and three letters that make up the angle. The angles for the illustration are ∠*COB*, ∠*COA*, ∠*AOD*, and ∠*BOD*. Note that the middle letter of each angle is always the vertex of the angle. Angles may also be indicated by a lowercase letter inside an angle (∠). Angles may also be formed by rotating one line from another line, provided they have a common endpoint. These lines are called *rays*. By definition, a ray is part of a line that begins at a given point and goes in one direction without ending. The symbol for a ray is $\overrightarrow{JK}$. See Figure 11-11. Angle *LJK* is formed by

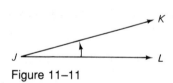

Figure 11-11

rotating ray *JK* in a counterclockwise direction from ray *JL*. Unless otherwise indicated, angles are generated in a counterclockwise direction as indicated by the curved arrow between the rays *JL* and *JK*. If a ray is rotated completely around another ray from a common vertex, it will make a complete revolution. A complete revolution is divided into 360 parts called *degrees*. See Figure 11-12. When ray *JK* is rotated completely around

Figure 11-12

from ray *JL* as indicated by the curved arrow, it has made a complete revolution of 360°. Degrees are indicated by the symbol °. Thus 35 degrees is written as 35°. For more accurate measurement, each degree is divided into 60 parts called *minutes* and is indicated by the symbol '. Each minute is divided into 60 parts, called *seconds,* and is indicated by the symbol ". When a line or ray rotates 180° or half of a complete revolution, it forms a *straight angle*. See Figure 11-13. When a line

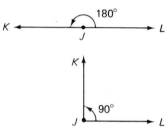

Figure 11-13

or ray rotates 90° or a quarter of a complete revolution, it forms a *right angle*. The symbol for a right angle is ∟. Lines or rays that form a right angle are said to be *perpendicular* to each other. The symbol for perpendicular is ⊥.

An angle greater than 90° and less than 180° is called an *obtuse angle*. An angle less than 90° is called an *acute angle*. Two angles whose sum is 90° are called *complementary angles*.

A simple illustration of angles is the hands of a clock. Each hour, the minute hand of a clock moves 360°. If the point where the hands of the clock are attached is considered to be the vertex of an angle and the hands are considered as sides or rays of the angle, an infinite number of angles would be formed. Since a clock has 12 numbers, the degrees between numbers, $\frac{360°}{12} = 30°$.

If a clock has indications for minutes, the degrees between each minute would be $\frac{360°}{60} = 6°$.

Example 11-2

PROBLEM: What angle is formed by the hands of a clock at 4 o'clock?

SOLUTION: Draw a sketch of a clock indicating 4 o'clock. See Figure 11–14. First, count the number of hours between 12 o'clock and 4 o'clock. The number is 4.

Second, multiply the number of hours (4) times the degrees between numbers on the face of the clock (30°) to arrive at the angle indicated by the hands of the clock (120°). 4 × 30° = 120°.

Figure 11–14

EXERCISE 11–2

Complete the following exercises concerning geometric definitions.

11–11. Sketch and label two intersecting line segments and their point of intersecton.

11–12. Sketch three lines and have them intersect at three different points. Label the points of intersection A, B, and C.

11–13. How many degrees are there in a half revolution?

11–14. How many degrees are there in the complement of a 68° angle?

11–15. How many degrees are there in the supplement of a 58° angle?

11–16. The complement of a 36° angle contains how many degrees?

11–17. How many degrees are there in the supplement of a 118° angle?

11–18. How many degrees are there in an acute angle?

11–19. How many degrees are there in a straight angle?

11–20. How many degrees are there in a right angle?

11–21. How many degrees are in a fourth of a right angle?

11–22. The number of degrees in a circle is equal to _____ straight angles and _____ right angles.

11–23. How many degrees are there in an obtuse angle?

11–24. Name the angle shown in Figure 11–15.

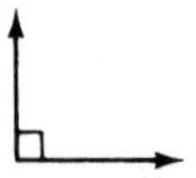

Figure 11–15

11–25. What type of angle is shown in Figure 11–16?

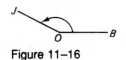

Figure 11–16

11–26. What type of angle is shown in Figure 11–17?

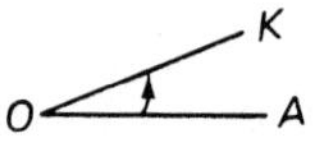

K
O   A

Figure 11–17

11–27. What angle will the hands of a clock form at 2 o'clock?

11–28. What central angle will the hands of a clock form at 5 o'clock?

11–29. What obtuse angle will the hands of a clock form at 8 o'clock?

11–30. If you were driving due north and made a 90° turn to the right, what would be your new direction?

MEASURING ANGLES

Angles are constructed and measured using a *protractor*. Study the protractor illustrated in Figure 11–18. The protractor may be divided into 10° and 1° increments.

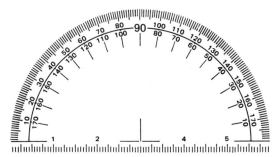
Figure 11–18

Example 11–3

PROBLEM: Measure a given acute angle using a protractor.

SOLUTION: First, place the protractor on the acute angle as illustrated in Figure 11–19. The flat edge of the protractor should be placed on ray *JL*, with the midpoint of the flat edge directly on point *J*. Second, read the inside degrees from right to left. Thus *KJL* is 52°.

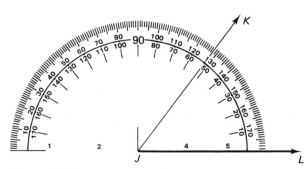
Figure 11–19

Example 11–4

PROBLEM: Measure a given obtuse angle using a protractor.

SOLUTION: First, place the protractor on the obtuse angle as illustrated in Figure 11–20. The flat edge of the protractor should be placed on ray *CB*, with the midpoint of the flat edge directly on point *C*. Second, read the inside degrees from right to left. Thus ∠*KJL* is 133°.

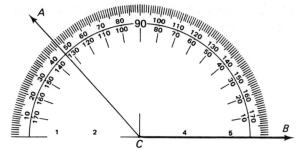
Figure 11–20

Example 11–5

PROBLEM: Construct an angle of 37°.

SOLUTION: First, draw a "working" ray, *FG*, and assign the vertex of the angle as point *F* on working ray *FG* as shown in Figure 11–21. Second, place the protractor on ray *FG* with the midpoint of the flat edge directly at point *F*. Third, read the inside degrees from right to left and make a point at 37°. Fourth, draw in ray *FH* to form ∠*HFG*.

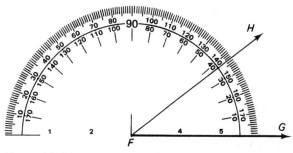
Figure 11–21

Example 11–6

PROBLEM: Construct an angle of 162°.

SOLUTION: First, draw, a working ray, *ST*, and assign the vertex of the angle at point *S* on working ray *ST* as shown in Figure 11–22. Second, place the

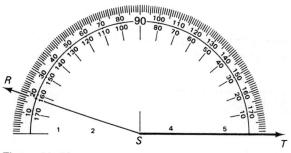
Figure 11–22

Chapter 11 / Basic Definitions and Properties of Geometry

protractor on ray *ST* with the midpoint of the flat edge directly at point *S*. Third, read the inside degrees from right to left and make a point at 162°. Fourth, draw in ray *SR* to form ∠*RST*.

EXERCISE 11–3

Using a protractor, measure the degrees of the angles and draw the angles as indicated.

11–31. In Figure 11–23, ∠*AOB* = _____.

11–32. In Figure 11–23, ∠*BOD* = _____.

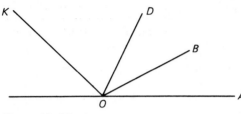

Figure 11–23

11–33. In Figure 11–23, ∠*AOK* = _____.

11–34. In Figure 11–24, ∠*RST* = _____.

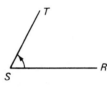

Figure 11–24

11–35. In Figure 11–25, ∠*PQR* = _____.

11–36. In Figure 11–26, ∠*XYZ* = _____.

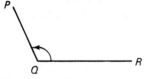

Figure 11–25

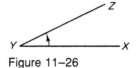

Figure 11–26

11–37. Construct ∠*KOB* as 48° in Figure 11–27.

11–38. Construct ∠*ROT* as 163° on Figure 11–28.

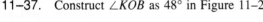

Figure 11–27

Figure 11–28

11–40. In Figure 11–29, ∠*AOC* = _____°.

11–39. Constuct the supplement of a 51° angle.

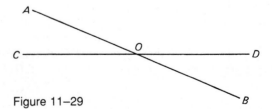

Figure 11–29

THINK TIME

The number of possible geometric figures is infinite. The names for geometric figures are determined by the number of sides or size of angles. Sometimes the figure is named by the number of sides used to construct a planed closed figure. A *polygon* is a geometric figure that has three or more sides. A *pentagon* is a five-sided figure. A *hexagon* is a six-sided figure. A *septagon* is a seven-sided figure. An *octagon* is an eight-sided figure. If all sides and angles are equal, the polygon is called a *regular polygon*. For example, a square is a regular polygon.

Some of the more common figures and their descriptions have been indicated in the following chart. Complete the chart to use as a reference.

	Name	*Figure*	*Number of Sides*	*Special Properties*
1.				Equal-length sides, four 90° angles
2.	rectangle			
3.				
4.				
5.	parallelogram			Both pairs of sides are parallel
6.	rhombus			Equal-length sides; Opposite angles equal
7.			5	
8.				
9.				
10.	circle			Curved line in which all points are equally distant from center

Chapter 11 / Basic Definitions and Properties of Geometry

PROCEDURES TO REMEMBER

1. To measure an angle, place the flat edge of a protractor on one of the rays or sides of the angle so that the midpoint of the flat edge of the protractor is directly on the vertex. Read the inside degrees of the protractor for angles less than 90°. Read the outside degrees for angles greater than 90°.
2. To construct an angle, draw a working ray and vertex, and place the flat edge of the protractor on the working ray or side with the midpoint of the flat edge of the protractor on the vertex. Read the degrees and make a mark at the given number of degrees. Draw a line from the vertex through the mark to form the angle.

CHAPTER SUMMARY

1. The following geometric terms and definitions are essential in solving geometry problems.
   (a) *Point* is a position or location that has no size or dimension.
   (b) A *line* is a series or set of points which, when connected, extend indefinitely in opposite directions or may be arranged to form a straight, broken, or curved line.
   (c) A *straight line* is the shortest distance between two points.
   (d) A *plane* is a flat surface that has length and width but no depth or height.
   (e) *Plane geometry* involves two dimensions: length and width.
   (f) *Solid geometry* involves three dimensions: length, width, and height.
   (g) A *plane figure* is any figure that can be drawn on a flat surface.
   (h) A *plane closed figure* is any plane figure that encloses an area.
   (i) A *polygon* is a plane closed figure described by straight-line segments and angles.
   (j) A *triangle* is a polygon with three straight line segments and angles.
   (k) A *right triangle* is a triangle that contains a right angle.
   (l) An *equilateral triangle* is a triangle with three equal sides.
   (m) An *isosceles triangle* is a triangle with two equal sides.
   (n) A *parallelogram* is a four-sided plane with opposite sides parallel.
   (o) A *rectangle* is a parallelogram with all right angles.
   (p) A *square* is a parallelogram with equal sides and all right angles.
   (q) A *rhombus* is a parallelogram with equal sides.
   (r) A *rectangular solid* is a geometric figure with length, width, and height; the six face planes are rectangular.
   (s) A *line segment* is a portion of a line indicated by endpoints.
   (t) A *vertex* of a solid is the point of intersection of three planes.
   (u) A *ray* is a part of a line that begins at a given point and goes in one direction.
   (v) A *degree* is a unit of angular measure equal to $\frac{1}{360}$ of a revolution. There are 360° in one complete revolution.
   (w) A *straight angle* contains 180°.
   (x) A *right angle* contains 90°.
   (y) An *obtuse angle* is an angle greater than 90° and less than 180°.
   (z) *Complementary angles* are two angles whose sum is 90°.
   (aa) An *acute angle* is an angle less than 90°.
   (bb) *Supplementary angles* are two angles whose sum is 180°.
   (cc) A *protractor* is an instrument used to measure and construct angles.
2. The following geometric symbols are important for use in working problems in geometry.
   (a) Curved line a: $\overset{\frown}{a}$
   (b) Broken line b: $\wedge\!\!\wedge\!\!\wedge_b$
   (c) Straight line c: $\overleftrightarrow{c}$ or $\overline{c}$
   (d) Point D: . D
   (e) Parallel lines: $\parallel$
   (f) Line segment AB: $\overline{AB}$
   (g) Angle ABC: $\angle ABC$ or $\angle B$
   (h) Ray JK: $\overrightarrow{JK}$
   (i) 276 degrees: 276°
   (j) 53 minutes: 53'
   (k) 16 seconds: 16"
   (l) A right angle: ∟
   (m) Perpendicular lines: ⊥

CHAPTER TEST

T–11–1. The symbol ⌐ means _____ angle.

T–11–2. The symbol $\overrightarrow{AB}$ means _____.

T–11–3. Write the symbol for line segment *LK*.

T–11–4. Write the symbol for line *n*.

T–11–5. How long is a ray?

T–11–6. If you were flying south and made a 90° turn to the right, in what direction would you be going?

T–11–7. The symbol for perpendicular lines is _____.

T–11–8. An instrument used to measure degrees is called a _____.

T–11–9. Degrees used to measure angles are subdivided into _____.

T–11–10. Angles are labeled by three capital letters. The middle letter indicates the _____ of the angle.

T–11–11. A triangle with two equal length sides is a (an) _____ triangle.

T–11–12. What degree of angle would the hands of a clock make at 5 o'clock?

T–11–13. Measure with a protractor angle *r* shown in Figure 11–30.

T–11–14. Measure with a protractor angle *PQR* shown in Figure 11–31.

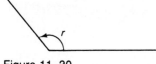

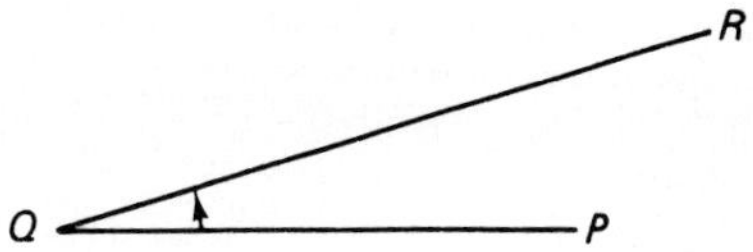

Figure 11–30

Figure 11–31

T-11-15. Construct a 39° angle with a protractor.

T-11-16. Construct a 121° angle with a protractor.

T-11-17. What is the complement of a 26° angle?

T-11-18. Determine the supplement of a 46° angle.

T-11-19. How many degrees are there in one third of a straight angle?

T-11-20. A position or location that has no size or dimension is called _____.

T-11-21. The point of intersection of three planes or angles is called _____.

T-11-22. A triangle with equal-length sides is called _____.

T-11-23. Measure ∠JOL in Figure 11-32.

T-11-24. Measure ∠MOJ in Figure 11-32.

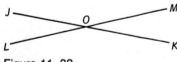

Figure 11-32

T-11-25. Angles *MOJ* and *JOL* are called _____ angles.

12

PERIMETERS AND AREAS OF PLANE GEOMETRIC FIGURES

OBJECTIVES

1. To develop the skills necessary to find the perimeter, circumference, and area of plane geometric figures.

2. To solve applied problems of industry and daily living that include perimeter, circumference, and area of geometric figures.

SELF-TEST

The following test will measure your ability to find the perimeter, circumference, or area of a plane geometric figure. Your successful completion of these problems shows that you know how to find perimeters and areas and you are ready to complete more advanced geometry problems. Problems that you are unable to solve will show areas you need to study.

S–12–1. What type of geometric figure is Figure 12–1?

S–12–2. Find the perimeter of Figure 12–1.

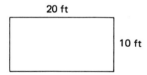

Figure 12–1

S–12–3. Find the area of Figure 12–1.

S–12–4. Figure 12–2 is what type of geometric figure?

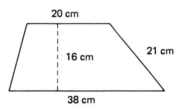

Figure 12–2

S–12–5. Find the perimeter of Figure 12–2.

S–12–6. Determine the area of Figure 12–2.

179

S-12-7. What type of geometric figure is Figure 12-3?

S-12-8. Find the perimeter of Figure 12-3.

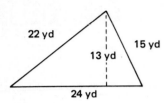

Figure 12-3

S-12-9. What is the altitude of Figure 12-3?

S-12-10. Calculate the area of Figure 12-3.

S-12-11. What is the diameter of Figure 12-4?

S-12-12. Find the area of Figure 12-4.

Figure 12-4

S-12-13. Find the circumference of Figure 12-4.

S-12-14. Find the amount of fencing needed to fence the garden area represented by Figure 12-5.

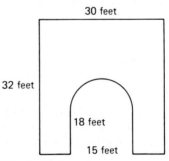

Figure 12-5

S-12-15. Find the area of the garden shown in Figure 12-5.

S-12-16. Find the inside circumference of the sidewalk around the fountain illustrated by Figure 12-6.

Figure 12-6

Section Three / Applied Geometry

S–12–17. What is the outside circumference of the sidewalk shown in Figure 12–6?

S–12–18. Find the area of the sidewalk shown in Figure 12–6.

S–12–19. Calculate the perimeter of an ice-skating rink represented by Figure 12–7.

S–12–20. What is the area of the ice-skating rink shown in Figure 12–7?

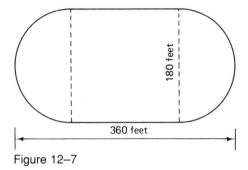

Figure 12–7

INTRODUCTION

The use of perimeters, circumferences, and areas of geometric figures is a daily experience for many people. However, geometric principles are seldom considered when a slice of bread is spread with peanut butter, a pie crust is rolled out, a wall is painted, or a tire size is selected. Yet in each case geometric problems are solved either directly or indirectly.

Formulas used to find perimeters, circumferences, and areas can be useful in saving both time and money at home and work. Knowing how to use formulas to find areas is of great value when, for example, carpeting a room or buying a lot. Determining perimeters, circumference, and area is necessary for many crafts and tradespeople.

DEFINITIONS

Perimeter is the distance around a geometric figure or the sum of the length of the sides. It is stated in linear measure such as inches, feet, miles, centimeters, and kilometers. The perimeter of a circle is called the *circumference* of the circle. *Area* is the amount of surface of a geometric figure.

The procedures for finding perimeter, circumference, and area can best be illustrated through the use of examples. Study each of the examples, noting the formulas used and the procedures followed.

SQUARES, RECTANGLES, AND TRIANGLES

Example 12–1

PROBLEM: Find the perimeter of the square shown in Figure 12–8.

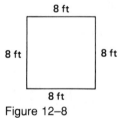

Figure 12–8

SOLUTION: The perimeter of the square can be found by adding the lengths of the four sides. Thus

$$\text{perimeter} = 8 \text{ ft} + 8 \text{ ft} + 8 \text{ ft} + 8 \text{ ft}$$
$$= 32 \text{ ft}$$

The formula for determining the perimeter of a square is

$$P = s + s + s + s \quad \text{or} \quad P = 4s$$

Chapter 12 / Perimeters and Areas of Plane Geometric Figures

Example 12–2

PROBLEM: Find the perimeter of the rectangle shown in Figure 12–9.

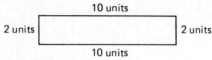

Figure 12–9

SOLUTION: The perimeter of the rectangle is the sum of the lengths of the four sides. Thus

$$\text{perimeter} = 10 \text{ ft} + 2 \text{ ft} + 10 \text{ ft} + 2 \text{ ft}$$
$$= 24 \text{ ft}$$

The formula for determining the perimeter of a rectangle is

$$\text{perimeter} = \text{length} + \text{width} + \text{length} + \text{width}$$
$$P = l + w + l + w$$
$$P = 2l + 2w$$
$$P = 2(l + w)$$

Example 12–3

PROBLEM: Find the perimeter of the triangle shown in Figure 12–10.

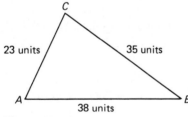

Figure 12–10

SOLUTION: The perimeter of the triangle is the sum of the lengths of the three sides. Thus

$$\text{perimeter} = 23 \text{ units} + 35 \text{ units} + 38 \text{ units}$$
$$P = 96 \text{ units}$$

The formula for determining the perimeter of a triangle is

$$\text{perimeter of a triangle} = \text{side } a + \text{side } b + \text{side } c$$
$$P = a + b + c$$

Example 12–4

PROBLEM: Find the area of the shown in Figure 12–11.

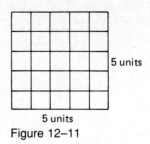

Figure 12–11

SOLUTION: One way to find the area of a square is to count each of the individual smaller squares. An easier way is to count the number of units in one column or side and then multiply this number by the number of rows. Thus

$$\text{area} = 5 \text{ units} \times 5 \text{ units}$$
$$= 25 \text{ square units}$$

The formula for determining the area of a square is

$$\text{area} = \text{side} \times \text{side}$$
$$A = s^2$$

It is important that area always be calculated in like units; thus the area should be in square inches, square yards, square meters, and so on.

Example 12–5

PROBLEM: Find the area of the rectangle shown in Figure 12–12.

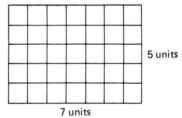

Figure 12–12

SOLUTION: The area of a rectangle can be found by multiplying the number of square units in one row (7) by the number of columns (5). Thus

$$\text{area} = 7 \text{ units} \times 5 \text{ units}$$
$$A = 35 \text{ square units}$$

The formula for finding the area of a rectangle is

$$\text{area} = \text{length} \times \text{width}$$
$$A = lw$$

Example 12-6

PROBLEM: Find the area of the parallelogram shown in Figure 12-13.

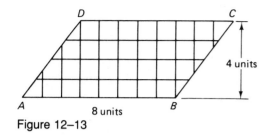

Figure 12-13

SOLUTION: One way to determine the area of the parallelogram is to move triangle ADE to the right side BC to form rectangle DE(E)C (see Figure 12-14) and

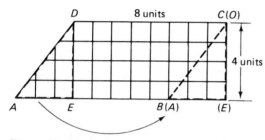

Figure 12-14

then multiply length DC times width DE. An easier way is to multiply the length DC times the height DE. Thus

$$\text{area} = 8 \text{ units} \times 4 \text{ units}$$
$$= 32 \text{ square units}$$

The formula for finding the area of a parallelogram is

$$\text{area} = \text{base (length)} \times \text{height (altitude)}$$
$$A = bh$$

Example 12-7

PROBLEM: Find the area of the triangle shown in Figure 12-15.

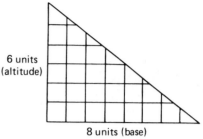

Figure 12-15

SOLUTION: One way to find the area of the triangle is to form a rectangle as illustrated in Figure 12-16.

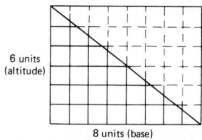

Figure 12-16

As shown in the sketch, the triangle is half of the area of the rectangle. Thus

$$\text{area} = \frac{8 \text{ units (base)} \times 6 \text{ units (altitude)}}{2}$$
$$= 24 \text{ square units}$$

The formula for determining the area of any triangle is

$$\text{area} = \frac{\text{base} \times \text{altitude}}{2}$$
$$A = \frac{ba}{2} \quad \text{or} \quad A = \frac{1}{2}bh \quad \text{or} \quad 0.5bh$$

EXERCISE 12-1

Solve the following problems for the given quantity.

12-1. Find the perimeter of the square shown in Figure 12-17.

12-2. Find the area of the square shown in Figure 12-17.

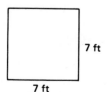

Figure 12-17

Chapter 12 / Perimeters and Areas of Plane Geometric Figures

12-3. Find the area of the square shown in Figure 12-18.

12-4. Find the perimeter of the square shown in Figure 12-18.

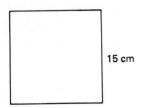

Figure 12-18

12-5. Find the area of the rectangle shown in Figure 12-19.

12-6. Find the perimeter of the rectangle shown in Figure 12-19.

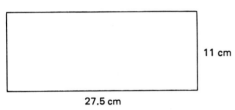

Figure 12-19

12-7. Find the perimeter of the rectangle shown in Figure 12-20.

12-8. Find the area of the rectangle shown in Figure 12-20.

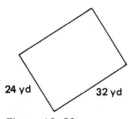

Figure 12-20

12-9. Calculate the area of the right triangle shown in Figure 12-21.

12-10. Find the perimeter of the right triangle shown in Figure 12-21.

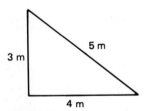

Figure 12-21

12–11. Find the area of the triangle shown in Figure 12–22.

12–12. Find the perimeter of the triangle shown in Figure 12–22.

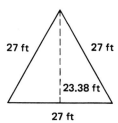
Figure 12–22

12–13. Find the perimeter of the triangle shown in Figure 12–23.

12–14. Find the area of the triangle shown in Figure 12–23.

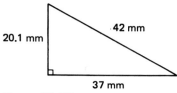
Figure 12–23

12–15. What is the altitude of the triangle shown in Figure 12–24?

12–16. What is the perimeter of the triangle shown in Figure 12–24?

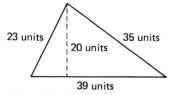
Figure 12–24

12–17. Find the area of the triangle shown in Figure 12–24.

12–18. What is the altitude of triangle shown in Figure 12–25?

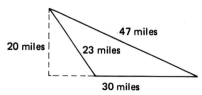
Figure 12–25

12–19. Find the perimeter of the triangle shown in Figure 12–25.

12–20. Determine the area of the triangle shown in Figure 12–25.

Chapter 12 / Perimeters and Areas of Plane Geometric Figures

CIRCLES

A *circle* is a plane curve with a set of points that are the same distance from a fixed point. The fixed point is called the *center* of the circle. The perimeter or distance around a circle is called the *circumference* of the circle. Figure 12-26 shows a circle, its center, and its circumference.

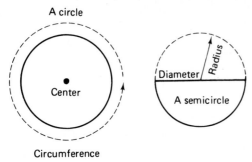

Figure 12-26

The *diameter* of a circle is a line drawn from one side of a circle, through the center, to the opposite side of the circle. The diameter divides the circle into two *semicircles*. Any line drawn from the center of a circle to the circumference of the circle is called a *radius*. The radius is half of the diameter or a circle. Study the illustration showing diameter, radius, and semicircle.

The *circumference* of a circle can be found by multiplying the diameter of the circle times 3.14. The special number, 3.14, is called *pi* (pronounced "pie") and its symbol is the Greek letter π. The approximate numerical value for π is 3.14. This number is the ratio of the circumference of a circle to its diameter. A more exact value for π is 3.141592653. For calculation purposes, π = 3.14 or 3.142 is used, depending on the accuracy desired. π was first determined by the ancient Egyptians as they developed a formula for finding the circumference of a circle. Most calculators have a π key, which gives π as 3.1415927.

The *area* of a circle can be determined by multiplying the radius of the circle × the radius of the circle × π.

Study the following examples for finding the circumference and area of a circle or semicircle.

Example 12-8

PROBLEM: Find the circumference of the circle shown in Figure 12-27.

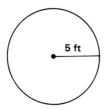

Figure 12-27

SOLUTION: The circumference of a circle is the distance around the circle. The circumference is found by multiplying the diameter (2 × the radius) by π, thus:

$$\text{circumference} = 2(5 \text{ ft}) \times 3.14$$
$$= 31.4 \text{ ft}$$

The formula for determining the circumference of a circle is

$$\text{circumference} = 2 \times \text{the radius} \times \pi$$
$$C = 2r\pi$$

or

$$\text{circumference} = \text{diameter} \times \pi$$
$$C = d\pi$$

Example 12-9

PROBLEM: Find the circumference of the semicircle shown in Figure 12-28.

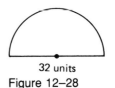

32 units

Figure 12-28

SOLUTION: The circumference of a semicircle is half the circumference of the whole circle plus the length of the diameter. First, determine the circumference of the whole circle.

$$C = d\pi$$
$$= 32 \text{ units} \times 3.14$$
$$= 100.48 \text{ units}$$

Second, divide the circumference (100.48 units) by 2.

$$\frac{100.48}{2} = 50.24 \text{ units}$$

Third, add the diameter (32 units) to half of the circumference (50.24 units).

$$50.24 \text{ units} + 32 \text{ units} = 82.24 \text{ units}$$

The formula for determining the circumference of a semicircle is

$$\text{circumference of a semicircle} = \frac{\pi \times \text{diameter}}{2} + \text{diameter}$$

$$C = \frac{\pi d}{2} + d$$

Example 12–10

PROBLEM: Find the area of the circle shown in Figure 12–29.

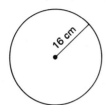

Figure 12–29

SOLUTION: The area of a circle is found by multiplying $\pi \times$ the radius $\times$ the radius. Thus

$$\text{area} = 3.14 \times 16 \text{ cm} \times 16 \text{ cm}$$
$$= 803.8 \text{ cm}^2$$

The formula for determining the area of a circle using the radius is

$$\text{area} = \pi \times \text{radius} \times \text{radius}$$
$$A = \pi r^2$$

Example 12–11

PROBLEM: Find the area of the circle shown in Figure 12–30.

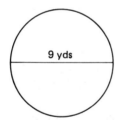

Figure 12–30

SOLUTION: The area of a circle, given the diameter, is determined by dividing the diameter by 2 to obtain the radius, and then multiplying $\pi \times$ the radius $\times$ the radius. First, determine the radius.

$$9 \text{ yd} \div 2 = 4.5 \text{ yd}$$

Second, use the formula for determining area using the radius.

$$\text{Area} = 3.14 \times 4.15 \text{ yd} \times 4.5 \text{ yd}$$
$$= 63.585 \text{ sq yd}$$

The formula for determining the area of a circle using the diameter is

$$\text{area} = \pi \times \left(\frac{\text{diameter}}{2}\right)^2$$

$$A = \pi \left(\frac{d}{2}\right)^2$$

$$= \frac{\pi d^2}{4}$$

This could be written $A = \dfrac{3.14159 d^2}{4}$, which could be written $A = 0.7854 d^2$, being that $3.14159 \div 4$ can be rounded to 0.7854. (Note where the numbers 0.7854 are found on a calculator.) Thus another formula to find the area of a circle is $A = 0.7854 d^2$.

Example 12–12

PROBLEM: Find the area of a 25-unit circle.

SOLUTION: The area of a circle, given the diameter, is found by multiplying 0.7854 by the diameter squared. A 25-unit circle means that the circle has a diameter of 25 units.

$$A = 0.7854 d^2$$
$$= 0.7854 \times 25 \text{ units} \times 25 \text{ units}$$
$$= 490.875 \text{ square units}$$

Example 12–13

PROBLEM: Find the area of the semicircle shown in Figure 12–31.

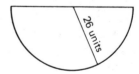

Figure 12–31

SOLUTION: The area of a semicircle is half the area of the whole circle. First, determine the area of the whole circle.

$$\text{Area} = 3.14 \times 26 \text{ units} \times 26 \text{ units}$$
$$\text{Area} = 2122.64 \text{ square units}$$

Second, divide the total area of the circle by 2 to determine the area of the semicircle.

$$\frac{1}{2} \text{ area of a circle} = \frac{2122.64}{2} \text{ square units}$$
$$= 1061.32 \text{ square units}$$

Chapter 12 / Perimeters and Areas of Plane Geometric Figures

The formula for determining the area of a semicircle is   or

$$\text{area of a semicircle} = \frac{\pi r^2}{2}$$

$$\text{area of a semicircle} = 0.5\pi r^2$$

EXERCISE 12-2

Solve the following problems for the given quantity.

12-21. Find the diameter of the circle shown in Figure 12-32.

12-22. Find the circumference of the circle shown in Figure 12-32.

Figure 12-32

12-23. Find the area of the circle shown in Figure 12-32.

12-24. Find the radius of the circle shown in Figure 12-33.

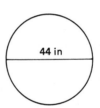

Figure 12-33

12-25. Determine the area of the circle shown in Figure 12-33.

12-26. Find the circumference of the circle shown in Figure 12-33.

12-27. Find the radius of the circle shown in Figure 12-34.

12-28. Find the circumference of the circle shown in Figure 12-34.

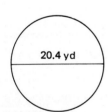

Figure 12-34

12–29. Find the area of the circle shown in Figure 12–34.

12–30. Find the diameter of the circle shown in Figure 12–35.

Figure 12–35

12–31. Calculate the area of the circle shown in Figure 12–35.

12–32. Find the circumference of the circle shown in Figure 12–35.

12–33. Determine the radius of the semicircle shown in Figure 12–36.

12–34. Find the circumference of the semicircle shown in Figure 12–36.

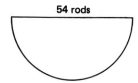

Figure 12–36

12–35. Find the area of the semicircle shown in Figure 12–36.

12–36. Determine the diameter of the semicircle shown in Figure 12–37.

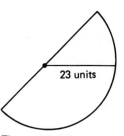

Figure 12–37

12–37. Find the area of the semicircle shown in Figure 12–37.

12–38. Find the circumference of the semicircle shown in Figure 12–37.

12–39. Find the area of a 7-foot circle.

12–40. What is the circumference of a 154.7-meter circle?

COMPLEX FIGURES

The basic formulas for finding the perimeter and area of a simple geometric figure can be used in more complex geometric figures. This is done by dividing the more complex figures into simple figures. The following examples illustrate how some of the changes are made and the formulas that are involved.

Example 12–14

PROBLEM: Find the area in Figure 12–38.

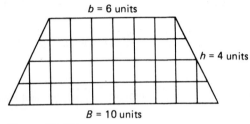

Figure 12–38

SOLUTION: The figure shown in Figure 12–38 is called a trapezoid. The upper base, b, and the lower base, B, are parallel to each other and the figure is closed. One way to determine the area of the trapezoid is to cut off the small triangles from each end of the trapezoid and reattach them to the upper corners to form a rectangle. See Figure 12–39. Note that the length of the newly formed rectangle is 8 units, or the average of the lengths of the upper base (6) and the lower base (10).

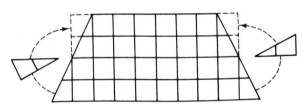

Figure 12–39

The height of the rectangle is 4 units. Thus

$$\text{area} = \frac{10 \text{ units} + 6 \text{ units}}{2} \times 4 \text{ units}$$
$$= \frac{16}{2} \text{ units} \times 4 \text{ units}$$
$$= 8 \text{ units} \times 4 \text{ units}$$
$$= 32 \text{ units}$$

The formula for determining the area of a trapezoid is

$$\text{area} = \frac{\text{upper base} + \text{lower base}}{2} \times \text{height}$$

or

$$A = \left(\frac{B + b}{2}\right)h$$

or

$$A = 0.5(B + b)h$$

Example 12–15

PROBLEM: Find the area in Figure 12–40.

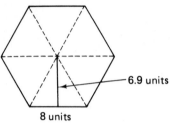

Figure 12–40

SOLUTION: The given figure is a hexagon. A *regular hexagon* is a polygon with six equal sides and angles, as shown in the sketch. Note that it can be divided into six equilateral triangles. Use the following procedures to determine the area of the given hexagon. First, find the area of one of the six equilateral triangles. Use the formula for finding the area of a triangle, $A = \frac{bh}{2}$.

$$A = \frac{8 \text{ units} \times 6.9 \text{ units}}{2}$$
$$= \frac{55.2 \text{ square units}}{2} = 27.6 \text{ square units}$$

Then multiply the area of one triangle (27.6 square units) by the number of triangles (6) to determine the area of the hexagon.

$$A = 27.6 \text{ square units} \times 6 = 165.6 \text{ square units}$$

The formula for determining the area of a hexagon is

$$\text{area} = 6\left(\frac{\text{base} \times \text{height}}{2}\right)$$

$$A = 6\left(\frac{bh}{2}\right) \quad \text{or} \quad 3bh$$

Example 12–16

PROBLEM: Find the cost of carpeting the floor plan shown in Figure 12–41. The cost of carpeting is $22 per yard.

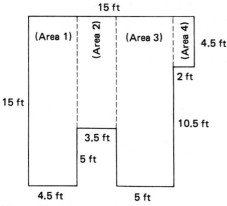

Figure 12–41

SOLUTION: First, divide the floor plan into rectangles as indicated by the dashed lines. Then find the area of each rectangle.

Area 1 = 15 ft × 4.5 ft = 67.5 sq ft
Area 2 = 10 ft × 3.5 ft = 35.0 sq ft
Area 3 = 15 ft × 5.0 ft = 75.0 sq ft
Area 4 = 4.5 ft × 2 ft = 9 sq ft

Find total the square footage of the four areas.

$$\begin{aligned} \text{Area 1} &= 67.5 \text{ sq ft} \\ \text{Area 2} &= 35.0 \text{ sq ft} \\ \text{Area 3} &= 75.0 \text{ sq ft} \\ +\text{ Area 4} &= 9.0 \text{ sq ft} \\ \hline \text{Total area} &= 186.5 \text{ sq ft} \end{aligned}$$

Since 9 sq ft = 1 sq yd, multiply as shown to convert sq ft to sq yd.

$$\left(\frac{186.5 \text{ sq ft}}{1}\right)\left(\frac{1 \text{ sq yd}}{9 \text{ sq ft}}\right) = 20.7 \text{ sq ft}$$

The amount of carpet needed is 20.7 square yards. Round off 20.7 square yards to the next full yard. Thus 20.7 square yards becomes 21 square yards needed. To find the cost of the carpeting, multiply the number of square yards (21) by the cost per square yard ($22).

$$\text{Cost} = \left(\frac{21 \text{ sq yd}}{1}\right)\left(\frac{\$22}{\text{sq yd}}\right) = \$462 \text{ for 21 sq yd}$$

Example 12–17

PROBLEM: Find the cost of bricks to build a wall 38 feet long and 4 feet high. Each brick with mortar covers an area of $8\frac{1}{2}$ inches by $2\frac{3}{4}$ inches and its cost is $0.38.

SOLUTION: Draw a sketch of the wall and the brick to gain a better understanding of what you are to do. See Figure 12–42.

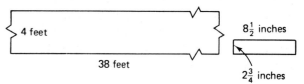

Figure 12–42

Convert all measurements to inches to eliminate possible confusion.

$$\left(\frac{38 \text{ ft}}{1}\right)\left(\frac{12 \text{ in.}}{1 \text{ ft}}\right) = 456 \text{ in.}$$

$$\left(\frac{4 \text{ ft}}{1}\right)\left(\frac{12 \text{ in.}}{1 \text{ ft}}\right) = 48 \text{ in.}$$

Find the area of the wall using the formula

$$\begin{aligned} A &= lw \\ &= (456 \text{ in.})(48 \text{ in.}) \\ &= 21{,}888 \text{ sq in.} \end{aligned}$$

Then convert the mixed fractions to decimal fractions to simplify the process.

$$8\frac{1}{2} \text{ in.} = 8.5 \text{ in.}$$

$$2\frac{3}{4} \text{ in.} = 2.75 \text{ in.}$$

Find the area of one brick using the formula

$$\begin{aligned} A &= lw \\ &= (8.5 \text{ in.})(2.75 \text{ in.}) \\ &= 23.375 \text{ sq in.} \end{aligned}$$

To find the number of bricks needed by dividing the area of the wall (21,888 square inches) by the area of one brick (23.375 square inches).

$$\left(\frac{21{,}888 \text{ sq in.}}{1}\right)\left(\frac{1 \text{ brick}}{23.375 \text{ sq in.}}\right) = 936.38 \text{ bricks}$$

Round the decimal fraction to the next complete whole number. Thus 936.36 bricks = 937 bricks. To find the total cost of the bricks, multiply the total number of bricks (937) by the cost per brick ($0.38).

$$\text{Cost} = \left(\frac{937 \text{ bricks}}{1}\right)\left(\frac{\$0.38}{1 \text{ brick}}\right) = \$356.06$$

This problem could be done more efficiently by using a calculator as follows:

cost of bricks = $\left(\dfrac{\text{total area of wall in inches}}{\text{area of one brick in inches}}\right)$ (cost per brick)

$= \left(\dfrac{38 \text{ ft}}{1} \times \dfrac{12 \text{ in.}}{1 \text{ ft}}\right)\left(\dfrac{4 \text{ ft}}{1} \times \dfrac{12 \text{ in.}}{1 \text{ ft}}\right)$

$\cdot \left(\dfrac{1 \text{ brick}}{(8.5 \text{ in.})(2.75 \text{ in.})}\right)$

$= \left(\dfrac{937 \text{ bricks}}{1}\right)\left(\dfrac{\$0.38}{1 \text{ brick}}\right)$

$= \$356.06$

EXERCISE 12–3

Solve the following problems concerning perimeters, circumferences, and areas to further develop your skills.

12–41. What kind of geometric figure is Figure 12–43?

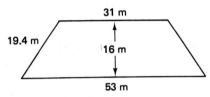

Figure 12–43

12–43. Find the area of Figure 12–43.

12–42. Find the perimeter of Figure 12–43.

12–44. Find the perimeter of Figure 12–44.

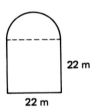

Figure 12–44

12–45. Find the area of Figure 12–44.

12–46. What kind of geometric figure is Figure 12–45?

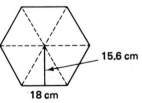

Figure 12–45

12–47. Find the perimeter of one of the triangles of Figure 12–45.

12–48. Find the perimeter of Figure 12–45.

12–49. Find the area of one of the triangles of Figure 12–45.

12–50. Calculate the area of Figure 12–45.

12–51. Find the area of Figure 12–46.

12–52. Find the perimeter of Figure 12–46.

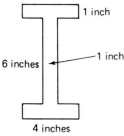

Figure 12–46

12–53. Find the perimeter of the triangle in Figure 12–47.

12–54. Find the perimeter of Figure 12–47.

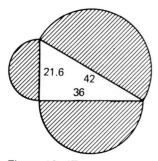

Figure 12–47

12–55. Find the area of the triangle in Figure 12–47.

12–56. Find the area of the shaded portion of Figure 12–47.

12–57. Calculate the area of Figure 12–47.

12–58. Find the perimeter of Figure 12–48.

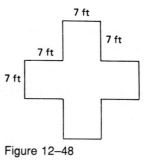

Figure 12–48

Chapter 12 / Perimeters and Areas of Plane Geometric Figures

12-59. Find the area of Figure 12-48.

12-60. What would be the cost of painting Figure 12-48 if the cost of paint and labor is $0.12 per square foot?

12-61. Calculate the area of Figure 12-49.

12-62. If Figure 12-49 represents a wheat field, how many bushels would be harvested per square acre if the total number of bushels for the entire area is 182? (160 square rods equals 1 acre.)

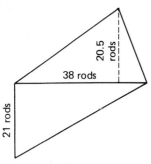

Figure 12-49

12-63. What is the cost of sodding the playing area of the football field represented by Figure 12-50 at the cost of $1.75 per square yard?

12-64. What is the cost of sodding the bowl ends of the field (Figure 12-50, areas B and C) at $1.15 per square yard?

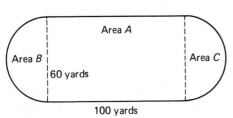

Figure 12-50

12-65. How many feet of fencing would be needed to fence in the entire area of Figure 12-50?

12-66. Find the cost of carpeting a 32-foot circular stage if the installed cost per yard is $34.95 per square yard.

12–67. How many feet of baseboard molding would be needed for the floor plan represented by Figure 12–51?

12–68. What is the total area represented by Figure 12–51?

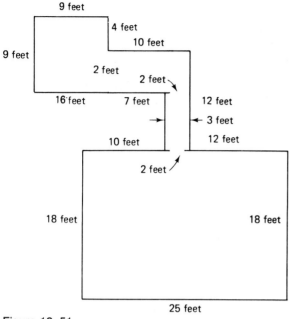

Figure 12–51

12–69. What is the cost of carpeting the rooms represented by Figure 12–51 at $28 per square yard? (1 square yard = 9 square feet.)

12–70. What is the wall area of a room represented by Figure 12–52?

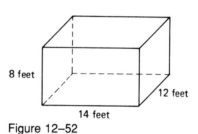

Figure 12–52

12–71. What is the area of the ceiling of the room shown in Figure 12–52?

12–72. Find the cost of installing hardwood parquet flooring in 6-inch squares at $0.89 per 6-inch square, shown in Figure 12–52.

12–73. How many gallons of paint are needed to paint the room shown in Figure 12–52 (walls and ceiling) if 1 gallon covers 550 square feet?

12–74. Using the information from problem 12–73, find the cost of painting the room (Figure 12–52) if the paint costs $19.95 per gallon and $5.95 per quart.

12–75. How many 2-inch by 2-inch tiles would be needed to cover the floor of a bathroom that is $7\frac{1}{2}$ feet by $8\frac{1}{2}$ feet?

12–76. A landscape contractor wishes to sod an area illustrated by Figure 12–53. How much sod, in sq. yds, will be needed?

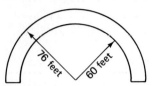

Figure 12–53

12–77. Find the cost of fencing Figure 12–53 if the cost is $1.29 per foot.

12–78. A carpenter has accepted the job of paneling a wall 12 feet by 72 feet. How many panels will he need if each panel is 4 feet by 8 feet?

12–79. A nurseryman finds that he needs 36 square feet of ground space to grow a spruce tree. How many spruce trees could he grow on a plot of land 200 yards by 300 yards?

12–80. What is the swimming area of a 35-foot-diameter pool?

THINK TIME

Formulas for finding the perimeter, circumference, or area of a plane geometric figure are difficult to remember. To refresh your memory from time to time, you may want to construct a chart for reference. Complete the chart below. When you finish, cover portions of the chart with a blank piece of paper to check your recall.

	Sketch	Name of Figure	Formula for Perimeter or Circumference	Formula for Area
1.	(square with sides s)			
2.		rectangle		
3.			$P = a + b + c$	
4.	(parallelogram with h, a, b)			
5.				$A = \pi r^2$
6.			$C = d + \dfrac{1}{2}\pi d$ or $C = 2r + \pi r$	
7.	(trapezoid)			
8.				$A = 6\left(\dfrac{bh}{2}\right)$

PROCEDURES TO REMEMBER

1. When determining the perimeter or the area, use only like units of measurement.

2. Convert mixed fractions to decimal fractions, where possible, when finding the area of a plane geometric figure.

3. Divide or change complex plane geometric figures into simple figures, where possible, and apply basic formulas for finding perimeter, circumference, or area.

Chapter 12 / Perimeters and Areas of Plane Geometric Figures

CHAPTER SUMMARY

1. The following definitions are essential to solving problems with plane geometric figures:
   (a) *Perimeter* is the distance around a plane geometric figure and is stated in linear measure such as inches, miles, centimeters, and meters.
   (b) *Circumference* is the distance around a circle.
   (c) *Area* is the amount of surface of a plane geometric figure and is measured in square inches, square yards, square meters, and so on.
   (d) A *circle* is a set of points that are the same distance from a fixed point called the *center*.
   (e) The *radius* of a circle is the distance from the center of the circle to the circumference.
   (f) The *diameter* of a circle is a line drawn from the circumference of a circle, through the center, to the opposite circumference.
   (g) A *semicircle* is half of a circle.
   (h) *Pi* is a special number, equal to approximately 3.14, which is the ratio for the circumference of a circle to its diameter. The symbol for pi is the Greek letter π. For calculation purposes, $\pi = 3.14$ or 3.142 is used, depending on the accuracy desired. The calculator value of π is 3.1415927.

2. The following formulas are essential to solving geometric figures.

Figure	Perimeter (Circumference)	Area
Square	$P = 4s$	$A = s^2$
Rectangle	$P = 2l + 2w$	$A = lw$
	$P = 2(l + w)$	
Triangle	$P = a + b + c$	$A = \dfrac{1}{2}bh$
Parallelogram	$P = 2a + 2b$	$A = bh$
	$P = 2(a + b)$	
Circle	$C = d\pi;$	$A = \pi r^2;$
	$C = 2\pi r$	$A = \dfrac{\pi d^2}{4}$
Semicircle	$C = d + \dfrac{1}{2}\pi d;$	$A = \dfrac{\pi r^2}{2}$
	$C = 2r + \pi r$	
Trapezoid	$P = 2a + B + b$	$A = \left(\dfrac{B + b}{2}\right)h$
Hexagon	$P = 6s$	$A = 6\left(\dfrac{bh}{2}\right)$

CHAPTER TEST

T–12–1. Figure 12–54 is what type of geometric figure?

814 km
Figure 12–54

T–12–2. Determine the area of Figure 12–54.

T–12–3. Find the perimeter of Figure 12–54.

T–12–4. What type of geometric figure is Figure 12–55?

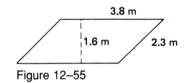

Figure 12–55

T–12–5. What is the altitude of Figure 12–55?

T–12–6. What is the perimeter of Figure 12–55?

T–12–7. Find the area of Figure 12–55.

T–12–8. What is the altitude of Figure 12–56?

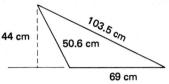
Figure 12–56

T–12–9. Find the perimeter of Figure 12–56.

T–12–10. What is the area of Figure 12–56?

T–12–11. What is the radius of Figure 12–57?

T–12–12. Calculate the area of Figure 12–57.

Figure 12–57

T–12–13. What is the circumference of Figure 12–57?

T–12–14. What kind of geometric figure is Figure 12–58?

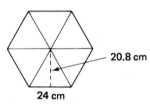
Figure 12–58

T–12–15. Find the perimeter of Figure 12–58.

T–12–16. Calculate the area of Figure 12–58.

T-12-17. Find the cost of carpeting the area around the swimming pool illustrated by Figure 12-59. The cost of carpeting is $18.95 per square yard. (9 square feet = 1 square yard.)

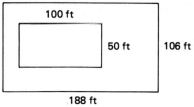
Figure 12-59

T-12-18. Two heating ducts are joined together as illustrated by Figure 12-60. Find what the diameter of the single opening (c) would have to be if the area of the openings is to be equal to the area of the openings $a + b$.

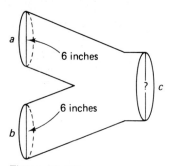
Figure 12-60

T-12-19. How many sections of acoustical ceiling tile will be needed to cover the ceiling illustrated by Figure 12-61? The tile sections are 2 feet by 4 feet.

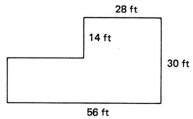
Figure 12-61

T-12-20. How many gallons of paint are needed to paint the walls and ceiling of the storage room illustrated by Figure 12-62? Each gallon of paint covers 520 square feet.

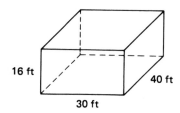
Figure 12-62

13

SURFACE AREAS AND VOLUMES OF GEOMETRIC FIGURES

OBJECTIVES

1. To develop skills to find lateral and total surface areas of prisms, pyramids, cylinders, cones, and spheres.
2. To develop skills to find volume of prisms, pyramids, cylinders, cones, and spheres.
3. To solve applied problems of industry and business that require finding areas and volumes of solid geometric figures.

SELF-TEST

This test will determine your ability to find lateral surface areas, total surface areas, and volumes of solid geometric figures. Successful completion of this self-test will indicate your ability to find areas and volumes of solid figures. Problems that you are unable to solve will indicate areas that will need further study.

S–13–1. Find the volume of the prism shown in Figure 13–1.

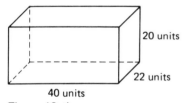

Figure 13–1

S–13–2. Find the total surface area of a 15-inch cube.

S–13–3. Find the lateral surface area of a rectangular solid 3 centimeters wide × 4 centimeters long × 30 centimeters high.

S–13–4. Find the lateral surface area of the cylinder shown in Figure 13–2.

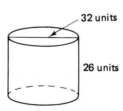

Figure 13–2

S–13–5. Find the volume of the cone shown in Figure 13–3.

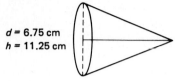

Figure 13–3

S–13–6. Find the volume of the cylinder shown in Figure 13–4.

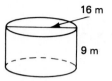

Figure 13–4

S–13–7. Find the total surface area of the cylinder shown in Figure 13–5.

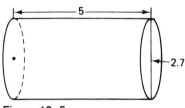

Figure 13–5

S–13–8. Find the lateral surface area of the cylinders shown in Figure 13–6.

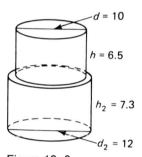

Figure 13–6

S–13–9. Find the volume of a 24-inch-diameter sphere.

S–13–10. Determine the volume of the pyramid shown in Figure 13–7.

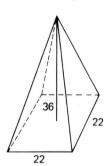

Figure 13–7

S–13–11. How many gallons will be held by the tank shown in Figure 13–8? (1 cubic foot = 7.5 gallons.)

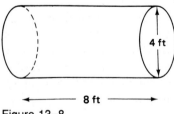

Figure 13–8

S–13–12. How many square feet of sheet metal are needed to construct the duct illustrated in Figure 13–9? (Add 15% for seams.)

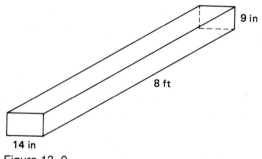

Figure 13–9

Section Three / Applied Geometry

S-13-13. Find the capacity of the cone portion of the funnel shown in Figure 13-10.

Figure 13-10

S-13-14. Find the cubic yards of concrete needed to construct the sidewalk shown in Figure 13-11.

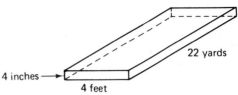

Figure 13-11

S-13-15. Determine the cost of painting a 50-foot-diameter spherical water tower at $1.89 per square yard.

INTRODUCTION

Every object, whether it is as large as the earth or as small as an atom, has surface area and volume. The surface area and volume of a geometric solid are determined by using formulas involving length, width, and height. Automobile manufacturers and buyers are concerned about "the cubic inches" or volume of compression of an engine. A smart shopper notes the volume of a container as well as its cost. An air-conditioning salesperson measures the volume of space to be air-conditioned. Many occupations require the skills to find the surface area and volume of geometric solids.

DEFINITIONS

The formulas for finding the surface area and volume of geometric solids are related to formulas used to find perimeter and area of plane figures. The skill with plane figure formulas will help find the areas and volumes of solid figures.

SURFACE AREA: PRISMS

A *prism* is a solid geometric figure with parallel edges and uniform cross sections. Figure 13-12 illustrates various prisms.

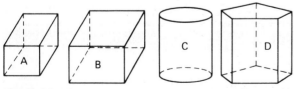

Figure 13-12

A *face* is any plane surface of a solid figure. A *lateral face* is a side of a geometric figure. The *bases* of a prism are the top face and bottom face, generally called the top and bottom. The *lateral edge* of a prism is where two sides of a solid intersect. Study Figure 13-13.

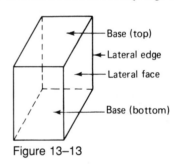

Figure 13-13

All prisms are classified by the shape of their base. A *right prism* has lateral faces that are perpendicular to the base. A prism whose sides are not perpendicular to its base is called an *oblique prism*.

Figure 13-12A is a *right square prism* because it has a square for a base. Figure 13-12B and 13-13 are *right rectangular prisms* because they have rectangular bases. Figure 13-12C represents a special type of prism called a *cylinder*. Figure 13-12D is a *right pentagonal prism* because of its pentagon-shaped base.

LATERAL SURFACE AREAS

The *lateral surface area* of a solid geometric figure is the area or surface of the sides of the figure, excluding

Chapter 13 / Surface Areas and Volumes of Geometric Figures

the area of the top and bottom. The shaded portion of Figure 13–14 indicates the lateral surface.

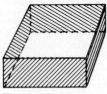

Figure 13–14

The *total surface area* of a geometric solid includes the area of the top and bottom of the figure as well as the area of the sides. To find the lateral surface of a prism, multiply the perimeter of the prism by its height.

Example 13–1

PROBLEM: Find the lateral surface (area) of Figure 13–15.

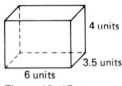

Figure 13–15

SOLUTION: First, determine the type of figure shown in Figure 13–15. Since the base of the figure is a rectangle and its faces are perpendicular to its base, Figure 13–15 is a right rectangular prism.

Then find the lateral area by finding the area of each of the four sides and adding these areas together. See Figure 13–16.

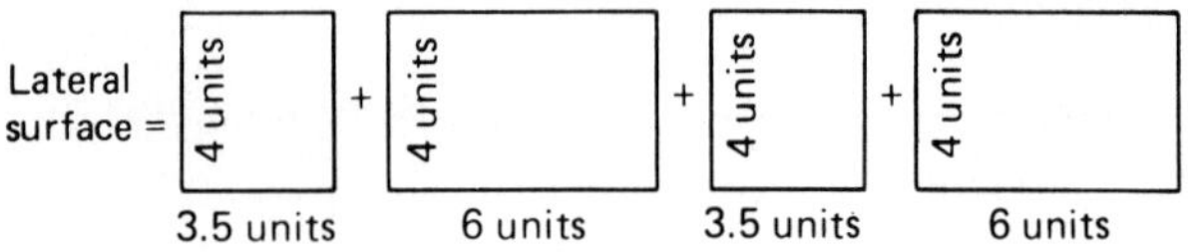

Lateral surface = [4 units × 3.5 units] + [4 units × 6 units] + [4 units × 3.5 units] + [4 units × 6 units]

Figure 13–16

$$\begin{aligned} \text{Lateral} \\ \text{surface} \end{aligned} = \begin{aligned} 4 \text{ units} \times 3.5 \text{ units} &= 24 \text{ square units} \\ 4 \text{ units} \times 6 \text{ units} &= 24 \text{ square units} \\ 4 \text{ units} \times 3.5 \text{ units} &= 24 \text{ square units} \\ 4 \text{ units} \times 6 \text{ units} &= 24 \text{ square units} \\ &= 76 \text{ square units} \end{aligned}$$

An easier way to find the lateral surface is to use the formula

$$\text{lateral surface (area)} = \text{perimeter } (2l + 2w) \times \text{height } (h)$$
$$A = p \times h$$
$$= [2(6) + 2(3.5)] \, 4$$
$$= 76 \text{ square units}$$

Example 13–2

PROBLEM: Find the total surface area of Figure 13–17.

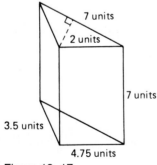

Figure 13–17

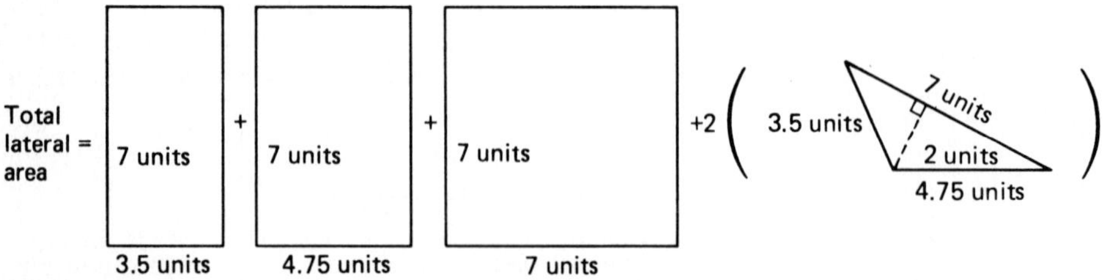

Total lateral area = [7 units × 3.5 units] + [7 units × 4.75 units] + [7 units × 7 units] + 2(triangle with 3.5 units, 7 units, 2 units, 4.75 units)

Figure 13–18

SOLUTION: Determine the type of figure shown in Figure 13–17. Since the base of the figure is a triangle and its faces are perpendicular to its base, the figure is a triangular prism.

Then find the surface areas. The total surface area may be found by finding the areas of each of the faces, the top, and the bottom and then adding these areas together. See Figure 13–18.

Total surface area =
$$\begin{aligned} 7 \text{ units} \times 3.5 \text{ units} &= 24.5 \text{ square units} \\ 4 \text{ units} \times 4.75 \text{ units} &= 33.25 \text{ square units} \\ 4 \text{ units} \times 7 \text{ units} &= 49 \text{ square units} \\ 2\left(\frac{1}{2}\right) \times 7 \text{ units} \times 2 \text{ units} &= 14 \text{ square units} \end{aligned}$$

Total surface area = 120.75 square units

The formula for determining the total surface area of a right triangular prism is

total surface area = perimeter $(2l + 2w)$
$\times$ height + $\dfrac{2 \text{ (base} \times \text{ height)}}{2}$

$A = p \times h + bh$

$A = (3.5 + 4.75 + 7)(7)$
$\quad + 2\,[0.5(7)(2)]$
$= (15.25)(7) + (2)(7)$
$= 106.75 + 14$
$= 120.75$ square units

EXERCISE 13–1

Find the lateral and total surface areas for the following as indicated.

13–1. What type of prism is Figure 13–19?

13–2. What is the lateral area of Figure 13–19?

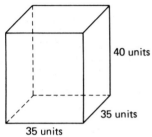
Figure 13–19

13–3. Find the total surface area of Figure 13–19.

13–4. What type of prism is Figure 13–20?

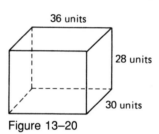
Figure 13–20

13-5. Find the lateral area of Figure 13–20.

13–6. Determine the total surface area of Figure 13–20.

13–7. What type of prism is Figure 13–21?

13–8. Find the lateral area of Figure 13–21.

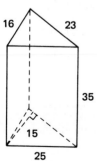
Figure 13–21

Chapter 13 / Surface Areas and Volumes of Geometric Figures

13–9. Determine the total surface area of Figure 13–21.

13–10. What type of prism is Figure 13–22?

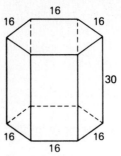

Figure 13–22

13–11. Find the lateral area of Figure 13–22.

13–12. Determine the total surface area of Figure 13–22.

SURFACE AREA: CYLINDERS AND CONES

Cylinders are commonly used in various ways in our daily lives. Many products are packaged in cylinders. Automobile engines are described by the number of cylinders. A *cylinder* is a solid geometric figure with equal, parallel, circular bases.

The *lateral surface area* of a cylinder is the surface area of the side of the cylinder, excluding the area of the top and bottom. The shaded portion of Figure 13–23 indicates the lateral surface area. The *total surface area* of a cylinder includes the area of the top and bottom plus the area of the side. The area of the shaded portion plus the top and bottom areas equal the total lateral surface of the cylinder, as shown by Figure 13–23.

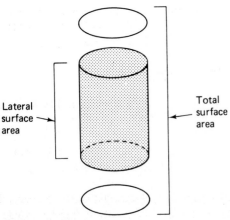

Figure 13–23

A *cone* is a solid geometric figure with one circular base whose side tapers evenly to a point. The *lateral surface area* of a cone is the surface area of the side of the cone, excluding the base area. The *total surface area* of a cone includes the area of the base and the area of the side. Figure 13–24 indicates the lateral surface. The lateral area plus the shaded portion of the base equals the total surface area of the cone.

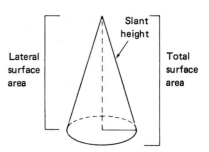

Figure 13–24

The *lateral surface area* of a cone is found by multiplying $\frac{1}{2}$ the diameter of the base (top or bottom) times π times the slant height. The *total surface area* of a cylinder is determined by multiplying $\frac{1}{2}$ the diameter of the base times π times the slant height plus the radius squared times π.

Example 13-3

PROBLEM: Determine the total surface area of the cylinder represented by Figure 13-25, with a diameter of 12 m and a height of 20 m.

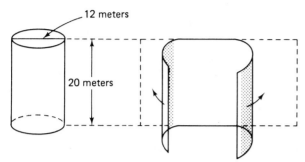

Figure 13-25

SOLUTION: Observe the cylinder; it could be "unrolled" to form a rectangle. If this were done, the lateral surface area of a cylinder could be found by multiplying the length times the width.

Then find the lateral surface area by multiplying the diameter (12 meters) × π (3.142) × the height (20 meters):

lateral surface area = 12 m × 3.142 × 20 m
 = 754.08 m²

Find the area of the bases (top and bottom). The bases are two equal circles; thus use the formula for finding the area of a circle. Multiply the radius × the radius × π (3.142) × 2, the number of bases. (Recall that the radius is $\frac{1}{2}$ of the diameter.)

Area of bases = $\frac{1}{2}$ (12 m) × $\frac{1}{2}$ (12 m) × 3.142 × 2
 = 226.224 m²

Then add the lateral surface area and the surface area of the bases to find the total surface area.

Total surface area = lateral surface area
 + area of bases
 = 754.08 m² + 226.224 m²
 = 980.304 m²

Note the formulas used to solve this problem:

Lateral surface area of a cylinder = $d\pi h$ or $2r\pi h$
Total surface area

of a cylinder = $d\pi h + 2 \left(\frac{d}{2}\right)^2 \pi$
 = $2r\pi h + 2\pi r^2$

Example 13-4

PROBLEM: Find the square feet of material needed to make the wind tunnel shown in Figure 13-26. The cone-shaped tunnel is 10 feet long and the radius is 2.5 feet.

SOLUTION: Observe the drawing of the cone in Figure 13-26. Note the height, h = 10 feet, the radius

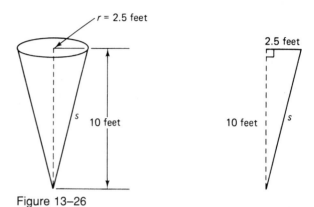

Figure 13-26

of the base r = 2.5 feet, and the slant height, s, is unknown. Find the slant height, s. Note in Figure 13-26 that a right triangle has been formed. Thus the formula for finding the hypotenuse of a right triangle can be used to find the slant height s.

Slant height = $\sqrt{(\text{radius})^2 + (\text{height})^2}$
 = $\sqrt{(2.5)^2 + (10)^2}$
 = $\sqrt{6.25 + 100}$
 = $\sqrt{106.25}$
 = 10.3 ft (to the nearest tenth)

Then find the lateral surface area of the cone.

Lateral surface area of the cone = lateral surface area = π
 = 2.5 ft × 10.3 ft × 3.142
 = 80.967 sq ft

Find the total surface area of the cone.

Total surface area of the cone = lateral surface area
 + πr^2
 = 80.967 sq
 + 19.6375 sq ft
 = 100.60 sq ft

Thus 101 square feet of material would be needed to make the wind tunnel cone. More material would be needed for seams and waste in the fabrication process.

Chapter 13 / Surface Areas and Volumes of Geometric Figures

The following formulas were used to solve this problem:

Lateral surface area of a cone = $r\pi \times$ slant height

Total surface area of a cone = $\pi \times$ slant height $+ \pi r^2$
= $\pi s + \pi r^2$

EXERCISE 13–2

13–13. Figure 13–27 is what type of figure?

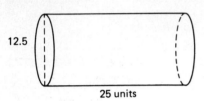

Figure 13–27

13–14. Find the lateral surface area of Figure 13–27.

13–15. Find the total surface area of Figure 13–27.

13–16. What type of solid is Figure 13–28?

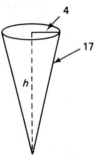

Figure 13–28

13–17. Find the diameter of Figure 13–28.

13–18. Find the height of Figure 13–28.

13–19. Find the lateral surface area of Figure 13–28.

13–20. Find the total surface of Figure 13–28.

13–21. What is the radius of Figure 13–29?

13–22. What is the slant height of Figure 13–29?

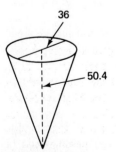

Figure 13–29

13-23. Determine the lateral surface area of Figure 13-29.

13-24. Find the total surface area of Figure 13-29.

SURFACE AREA: PYRAMIDS AND SPHERES

Another type of solid figure, similar to the cone, is the pyramid. The names of pyramids are determined by the shape of their base. A *right pyramid* is a pyramid with a regular polygon base whose sides form equal isosceles triangles. Study the drawings of three different types of pyramids: square pyramid (Figure 13-30), rectangular pyramid (Figure 13-31), and hexagonal pyramid (Figure 13-32); the first two are right pyramids. Note the names of the various parts of the pyramids.

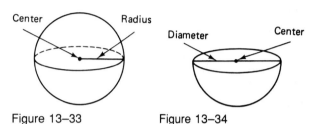

Figure 13-33 Figure 13-34

surface area and the base area. The *total surface area* of a sphere is found by the formula $4\pi r^2$ or πd^2. The *lateral surface area* of a hemisphere is found by the formula $2\pi r^2$ or $\dfrac{\pi d^2}{2}$. The *total surface area* of a hemisphere is determined by the formula $3\pi r^2$ or $0.75\ \pi d^2$.

Example 13-5

PROBLEM: Find the lateral and total surface areas of the pyramid shown in Figure 13-35.

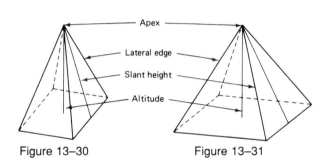

Figure 13-30   Figure 13-31

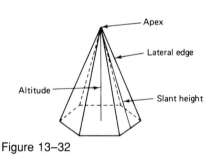

Figure 13-32

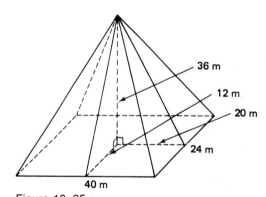

Figure 13-35

A *sphere* is a solid figure with all points on the surface the same distance from the center. A unique property of the sphere is that it has a greater volume in proportion to its surface area than that of any other solid geometric figure. A *hemisphere* is a half sphere. Study the sketches of the sphere (Figure 13-33) and hemisphere (Figure 13-34). Note the various parts of the sphere and hemisphere.

The *lateral surface area* of a pyramid is found by adding the area of each of the face triangles. The *total surface area* of a pyramid is found by adding the lateral

SOLUTION: Note that the figure is a rectangular pyramid because its base is a rectangle and all faces are isosceles triangles. Then find the slant height of two different faces. See Figure 13-36. Find the lateral surface area of the pyramid by using the formula for the surface area of a triangle. The lateral surface area of a pyramid is found by adding the surface areas of the four triangles.

Slant height
= $\sqrt{(12)^2 + (36)^2}$
= $\sqrt{144 + 1296}$
= $\sqrt{1440}$
= 37.9 meters

Slant height
= $\sqrt{(36)^2 + (20)^2}$
= $\sqrt{1296 + 400}$
= $\sqrt{1696}$
= 41.2 meters

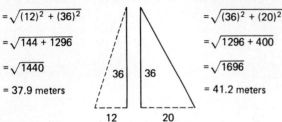

Figure 13–36

Lateral surface area
$= 2\left[\dfrac{\text{base (40 m)} \times \text{slant height (37.9 m)}}{2}\right]$
$+ 2\left[\dfrac{\text{base (24 m)} \times \text{slant height (41.2 m)}}{2}\right]$
$= \dfrac{1}{\cancel{2}}\left(\dfrac{40 \text{ m} \times 37.9 \text{ m}}{\cancel{2}}\right) + \dfrac{1}{\cancel{2}}\left(\dfrac{24 \text{ m} + 41.2 \text{ m}}{\cancel{2}}\right)$
$= 1516 \text{ m}^2 + 988.8 \text{ m}^2$
$= 2504.8 \text{ m}^2$

Then find the total surface area of the pyramid.

Total surface area = lateral surface area
 + base surface area
 $= 2504.8 \text{ m}^2 + (40 \text{ m})(24 \text{ m})$
 $= 2504.8 \text{ m}^2 + 960 \text{ m}^2$
 $= 3464.8 \text{ m}^2$

Note the formulas used to solve this problem:

Slant height $= \sqrt{(\text{altitude})^2 + \left(\dfrac{1}{2}\text{base}\right)^2}$

$\text{sh} = \sqrt{a^2 + \left(\dfrac{b}{2}\right)^2}$

Lateral surface area $= 2\left(\dfrac{1}{2}\text{base} \times \text{slant side 1}\right)$
$+ 2\left(\dfrac{1}{2}\text{base} \times \text{slant side 2}\right)$
$= 2\left(\dfrac{b}{2}\text{sh}_1\right) + 2\left(\dfrac{b}{2}\text{sh}_2\right)$
$= b(\text{sh}_1) + b(\text{sh}_2)$

Total surface area = lateral surface area + base area

Example 13–6

PROBLEM: Find the total surface area of the sphere shown in Figure 13–37.

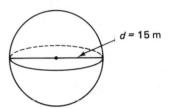

Figure 13–37

SOLUTION: The lateral surface area and the total surface area of a sphere are the same. Recall the formula for determining the total surface area of a sphere: $4\pi r^2$ or πd^2.

Total surface area $= \pi d^2$
 $= (3.142)(15 \text{ m})(15 \text{ m})$
 $= 706.95 \text{ m}^2$

Note the lateral surface area of a hemisphere is found by the formula $\dfrac{\pi d^2}{2}$, or $2\pi r^2$.

EXERCISE 13–3

Find the lateral and total surface areas as called for in the following problems.

13–25. What type of solid figure is Figure 13–38?

13–26. Find the slant height of Figure 13–38.

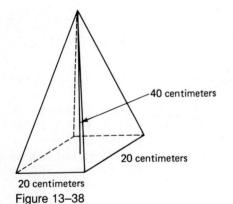

20 centimeters
Figure 13–38

13–27. Find the lateral surface area of Figure 13–38.

13–28. Find the total surface area of Figure 13–38.

13–29. What is the total surface area of Figure 13–39?

13–30. Determine the lateral surface area of the hemisphere of Figure 13–39.

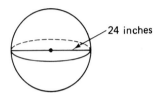

Figure 13–39

13–31. What kind of solid figure is Figure 13–40?

13–32. Find the slant height on the 45-unit side of Figure 13–40.

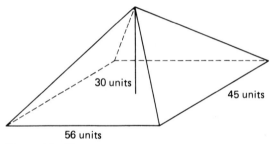

Figure 13–40

13–33. Find the slant height on the 56-unit side of Figure 13–40.

13–34. Find the lateral surface area of Figure 13–40.

13–35. Find the total surface area of Figure 13–40.

13–36. What type of figure is Figure 13–41?

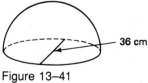

Figure 13–41

Chapter 13 / Surface Areas and Volumes of Geometric Figures

13-37. What is the radius of Figure 13-41?

13-38. What is the altitude of Figure 13-41?

13-39. Find the lateral surface area of Figure 13-41.

13-40. Determine the total surface area of Figure 13-41.

VOLUME

Volume is the amount of cubic space or content a solid figure contains. The volume of a prism or cylinder is found by multiplying the area of the base times the altitude or height.

To better understand volume, study Figure 13-42. The cubes in the first layer are numbered from 1 to 16. Recall that a cube is a solid figure with all square faces; thus all edges are equal. If each cube is 1 cubic inch, the volume of the first layer would be 16 cubic inches. The volume of all three layers would be 16 cubic inches per layer times the number of layers (3), or a total of 48 cubic inches.

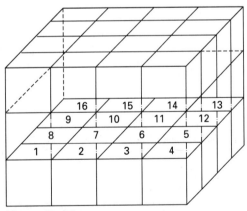

Figure 13-42

The volume of a pyramid or cone is found by multiplying the area of the base times the altitude or height times one third.

Example 13-7

PROBLEM: Find the volume of the rectangular solid shown in Figure 13-43.

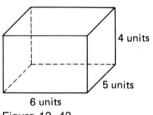

Figure 13-43

SOLUTION: Use the formula for finding the volume of a rectangular solid.

$$\text{Volume} = \text{length times width times height}$$
$$= (6 \text{ units})(5 \text{ units})(4 \text{ units})$$
$$= 120 \text{ cubic units}$$

Example 13-8

PROBLEM: Find the volume of the cylinder shown in Figure 13-44.

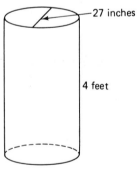

Figure 13-44

SOLUTION: First, convert to like units. (12 inches = 1 foot.)

$$\left(\frac{4 \text{ ft}}{1}\right)\left(\frac{12 \text{ in.}}{1 \text{ ft}}\right) = 48 \text{ in.}$$

Section Three / Applied Geometry

The radius $=\frac{1}{2}$ (27 inches). Then use the formula for finding the volume of a cylinder: $V = \pi r^2 h$.

Volume = area of base (πr^2) × height or altitude
= (3.142)(13.5 in.)(13.5 in.)(48 in.)
= 27,468.22 cu in.

Example 13–9

PROBLEM: Find the volume of the cone shown in Figure 13–45.

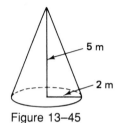

Figure 13–45

SOLUTION: Use the formula for finding the volume of a cone, $V = \frac{\pi r^2 h}{3}$.

$$\text{Volume} = \frac{\pi r^2 h}{3}$$
= area of base × altitude × one-third
$$= \frac{(3.142)(2)(2)(5)}{3}$$
$$= \frac{(3.142)(20 \text{ m}^3)}{3}$$
= 20.94 m³

The formula for the volume of a pyramid is

volume = area of the base × the height × one-third

Example 13–10

PROBLEM: Find the volume of an 8-inch sphere.

SOLUTION: Use the formula for finding the volume of a sphere.

Volume = 4 × π × the radius cubed ÷ 3
$$= \frac{(4)(3.142)(4)(4)(4)}{3}$$
= 268.12 in.³

The formula for determining the volume of a hemisphere is $V = \frac{2\pi r^3}{3}$.

EXERCISE 13–4

Find the volume of the following as indicated.

13–41. What type of solid is shown in Figure 13–46?

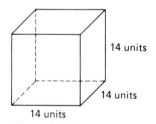

Figure 13–46

13–42. Find the volume of the first layer of the solid in Figure 13–46.

13–43. Find the volume of the solid in Figure 13–46.

13–44. What type of solid is Figure 13–47?

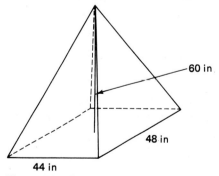

Figure 13–47

13-45. Find the area of the base of Figure 13-47.

13-46. Determine the volume of Figure 13-47.

13-47. What type of solid is shown in Figure 13-48?

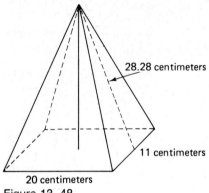

Figure 13-48

13-48. Find the altitude of Figure 13-48.

13-49. Find the area of the base of Figure 13-48.

13-50. Determine the volume of Figure 13-48.

13-51. What type of solid is shown in Figure 13-49?

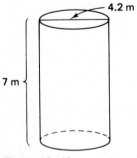

Figure 13-49

13-52. Find the area of the base of Figure 13-49.

13-53. Find the volume of Figure 13-49.

13-54. What type of solid is Figure 13-50?

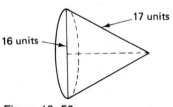

Figure 13-50

13–55. Find the area of the base of Figure 13–50.

13–56. Find the volume of Figure 13–50.

13–57. What type of solid is shown in Figure 13–51?

13–58. Find the volume of Figure 13–51.

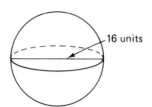

Figure 13–51

13–59. What type of solid is shown in Figure 13–52?

13–60. Find the volume of Figure 13–52.

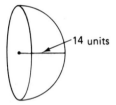

Figure 13–52

APPLICATIONS

The following examples illustrate applied problems. In each a formula for finding the volume of a solid is used. Sometimes more than one formula is used.

Example 13–11

PROBLEM: Find the capacity in gallons of the cylindrical tank shown in Figure 13–53.

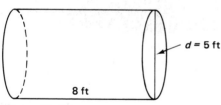

Figure 13–53

SOLUTION: Find the volume of the cylinder using the formula, $V = \pi r^2 h$.

$$V = (3.142)(2.5 \text{ ft})(2.5 \text{ ft})(8 \text{ ft})$$
$$= 157.1 \text{ cu ft}$$

Then find the number of gallons in 157.1 cubic feet using the fact that 1 cubic foot = 7.5 gallons.

$$\frac{157.1 \text{ cu ft}}{1} \times \frac{7.5 \text{ gal}}{1 \text{ cu ft}} = 1178.2 \text{ gal}$$

Example 13–12

PROBLEM: Find the tons of coal in an pile with a circumference of 62.8 feet and a height of 9 feet shown in Figure 13–54.

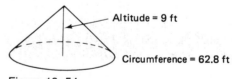

Figure 13–54

SOLUTION: Find the radius of the base of a pile. Use the formula $c = d\pi$; thus $\frac{c}{\pi} = d$.

$$\text{diameter} = \frac{\text{circumference}}{\pi}$$
$$= \frac{62.8 \text{ ft}}{\pi}$$
$$= 20 \text{ ft}$$
$$\text{radius} = 10 \text{ ft}$$

Chapter 13 / Surface Areas and Volumes of Geometric Figures

Then find the volume of the coal pile by using the formula for the volume of a cone.

$$V = \frac{\pi r^2 h}{3}$$
$$= \frac{(3.142)(10 \text{ ft})(10 \text{ ft})(9 \text{ ft})}{3}$$
$$= 942 \text{ cu ft}$$

To find the tons of coal by using the fact that 1 cubic foot = 0.0235 ton of coal.

$$\left(\frac{942 \text{ cu ft}}{1}\right)\left(\frac{0.0235 \text{ ton}}{\text{cu ft}}\right) = 222.137 \text{ tons of coal}$$

Example 13–13

PROBLEM: Find the amount of paint needed to paint the water tower shown in Figure 13–55.

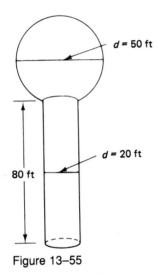

Figure 13–55

SOLUTION: Find the total surface area of the sphere using the formula πd^2.

Total surface area = (3.14)(50 ft)(50 ft)
 = 7859 sq ft

Then find the lateral surface area of the cylinder using the formula $LS = d\pi r h$.

Lateral surface area = (20 ft)(3.14)(80 ft)
 = 5024 sq ft

Determine the surface area that is under the sphere, inside the cylinder, where the sphere is attached to the cylinder, using the formula πr^2.

$$A = (3.14)(10 \text{ ft})(10 \text{ ft})$$
$$= 314 \text{ sq ft}$$

Determine the total surface area to be painted.

$$\begin{aligned}
\text{Total area of sphere} &= 7850 \text{ sq ft} \\
+ \text{ Total area of cylinder} &= 5024 \text{ sq ft} \\
\text{Total area} &= 12{,}874 \text{ sq ft} \\
- \text{ Area under the sphere} &= 314 \text{ sq ft} \\
\text{Area to be painted} &= 12{,}560 \text{ sq ft}
\end{aligned}$$

The gallons of paint needed: 1 gallon of paint covers 550 square feet.

$$\left(\frac{12{,}560 \text{ sq ft}}{1}\right)\left(\frac{1 \text{ gal}}{550 \text{ sq ft}}\right) = 22.84 \text{ gal or } 23 \text{ gal}$$

EXERCISE 13–5

Find the areas and volumes of the following applied problems.

13–61. How much fabric would be needed to cover a cylindrical wastebasket 18 inches high and 8 inches in diameter?

13–62. Find the amount of stainless steel needed to make a tank 6.75 meters long and 6 meters in diameter (Allow 5% for seams.)

13–63. Find the volume in cubic meters of the tank in problem 13–62.

13–64. Determine the weight of a brass solid 3 yards long, 1 inch thick, and $\frac{1}{2}$ foot wide. (Brass weighs 512 pounds per cubic foot.)

13–65. Find the capacity in gallons of a rectangular gas tank 12 inches by 14.5 inches by 48 inches. (1 gallon = 231 cubic inches.)

13–66. Find the capacity in gallons of a cylindrical gas tank 12 inches in diameter and 48 inches long. (1 gallon = 231 cubic inches.)

13–67. How many gallons of crude oil will a mile section of the Alaskan pipeline contain if the inside diameter is 48 inches? (1 cubic foot = 7.5 gallons.)

13–68. Find the capacity of the mill-dust collector shown in Figure 13–56.

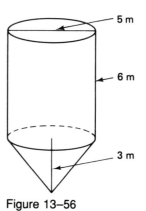

Figure 13–56

13–69. A stainless steel cylindrical milk tanker is 38 feet long and 5 feet in diameter. Determine the amount of stainless steel material needed to construct a tanker, allowing 12% for seams and waste.

13–70. Find the gallon capacity of the milk tanker in problem 13–69. (1 cubic foot = 7.5 gallons.)

13–71. A swimming pool is to be 50 meters long and 20 meters wide and will have an average depth of 3 meters. How many cubic meters must be excavated if an additional meter must be excavated for each side and the bottom?

13–72. Determine the square meters to be tiled if the inside and bottom of the pool (problem 13–71) are tiled.

13–73. A new house is built in the form of a hemisphere having a radius of 7 meters. If the air in the house is to be circulated every 90 seconds, how many cubic meters of air must be moved per hour?

13–74. A 50-foot length of $\frac{5}{8}$-inch garden hose contains how many quarts (to the nearest tenth)? (1 quart = 57.75 cubic inches.)

13–75. Find the monthly heating cost for a 6.5-foot-diameter circular hot tub filled to a depth of $2\frac{1}{2}$ feet. The cost to heat 1 gallon of water per month is $0.027. (1 cubic foot = 7.5 gallons.)

THINK TIME

To provide yourself with a simple review guide for geometric figures and the formulas for determining their areas, complete the following chart from the given sketches shown.

Sketch	Name of Figure	Formula for Lateral Surface Area	Formula for Total Surface Area	Formula for Volume
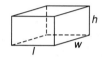				

PROCEDURES TO REMEMBER

1. Be certain to use only like units when finding an area or volume.
2. To find the area or volume of complex figures, change to simple figures where possible, and apply basic formulas for determining lateral surface area, total surface area, or volume.

CHAPTER SUMMARY

The following definitions are essential when working with geometric figures.

1. A *prism* is a solid geometric figure with parallel edges and uniform cross sections.
2. A *face* is any plane surface of a solid figure.
3. A *lateral face* of a prism is a side of a geometric figure.
4. The *bases* of a prism are the top face and the bottom face and are generally called the top and bottom of the solid.
5. The *lateral edge* of a prism is where two sides of a solid intersect.
6. A *right prism*'s faces are perpendicular to its base.
7. An *oblique prism*'s faces are not perpendicular to its base.
8. A *right square prism* has a square base.
9. A *right rectangular prism* has a rectangular base.
10. A *right pentagonal prism* has a pentagon-shaped base.
11. The *lateral surface area* of a solid geometric figure is the surface area of the sides not including the area of the top and bottom.
12. The *total surface area* of a solid geometric figure is the area of the top and bottom of the figure including the area of the sides.
13. A *right pyramid* is a pyramid with a regular polygon base and sides that form equal isosceles triangles.
14. A *sphere* is a solid figure with all points equal distances from the center.
15. A *hemisphere* is a half-sphere.
16. *Volume* is the amount of cubic units or content in a solid figure.

CHAPTER TEST

T–13–1. Find the volume of a rectangular solid 2 centimeters by 5 centimeters by 8 centimeters.

T–13–2. Find the lateral surface area of Figure 13–57.

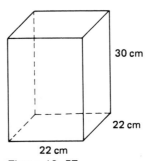

Figure 13–57

T–13–3. Determine the total surface area of a rectangular room 8 feet wide, 16 feet long, and 8 feet high.

T–13–4. Find the lateral surface area of Figure 13–58.

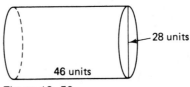

Figure 13–58

T–13–5. Find the volume in Figure 13–59.

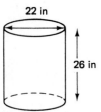

Figure 13–59

T–13–6. Find the lateral surface area of the cone in Figure 13–60.

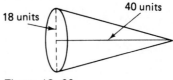

Figure 13–60

T-13-7. Determine the total surface area of the cylinders in Figure 13-61.

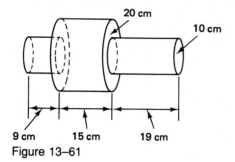

Figure 13-61

T-13-8. Determine the volume of the cone in Figure 13-62.

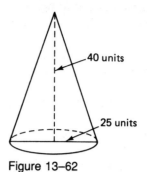

Figure 13-62

T-13-9. Find the surface area of the hemisphere in Figure 13-63.

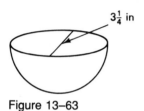

Figure 13-63

T-13-10. Determine the total surface area of the pyramid in Figure 13-64.

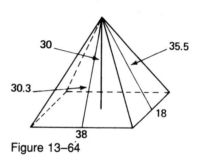

Figure 13-64

T-13-11. A stainless steel milk tanker is 36 feet long and 6 feet in diameter. How many gallons will it hold? (1 cubic foot = 7.5 gallons.)

T-13-12. Find the capacity of the sawdust collector shown in Figure 13-65.

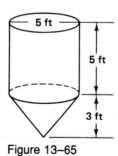

Figure 13-65

T-13-13. A cylindrical toothpaste container is 18 centimeters long and has a diameter of 3 centimeters. Find the number of cubic centimeters of toothpaste in the container.

T-13-14. A spherical crude-oil container has a diameter of 16 meters. How many liters of crude oil will the container hold if $1 \text{ m}^3 = 1000$ liters?

T-13-15. A pipeline in a refinery is 400 feet between shutoff valves. How many gallons of crude oil would be in the 400-foot section if the inner diameter of the pipe is 3 feet? (1 cubic foot = 7.5 gallons.)

14

CONSTRUCTION OF SIMPLE GEOMETRIC FIGURES

OBJECTIVE

1. To develop skills with a compass and straightedge to construct the following: circles, hexagons, squares, triangles, perpendiculars, parallel lines, angles (60°, 30°, 45°), tangents to circles, bisections, copy (lines, angles), and proportions.

SELF-TEST

This test will measure your ability to construct simple geometric forms. To complete this test, use a compass and straightedge. Construction of the figures will show your ability to construct geometric figures and your readiness to advance. Problems that you are unable to solve will indicate areas for improvement. When necessary, make your constructions on a separate sheet of paper.

S–14–1. Bisect $\overline{XY}$.

X———————Y

S–14–2. Bisect $\angle ABC$.

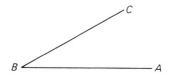

S–14–3. Copy $\angle FOB$.

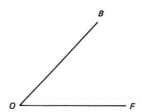

S–14–4. Divide $\overline{GJ}$ into three equal parts.

S–14–5. Construct a perpendicular on $\overline{JY}$ at point L.

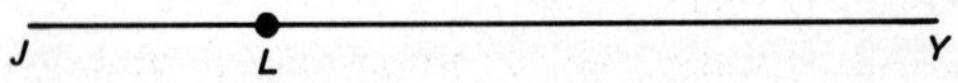

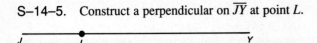

S–14–6. Construct a perpendicular to $\overline{KC}$ from point M.

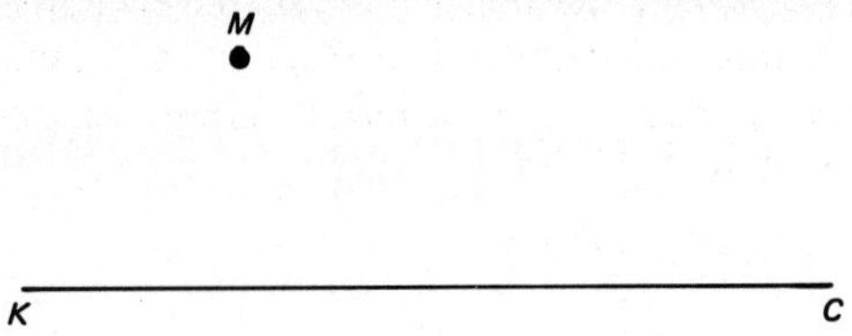

S–14–7. Construct a line parallel to $\overline{ND}$.

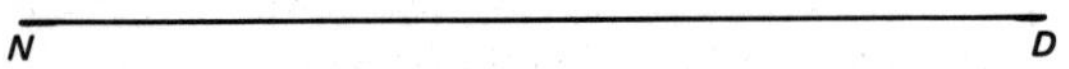

S–14–8. Construct tangents to the circle from point P.

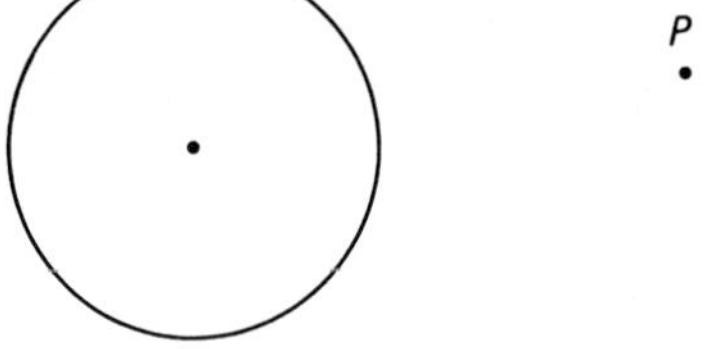

S–14–9. Construct a square with sides equal to $\overline{AB}$.

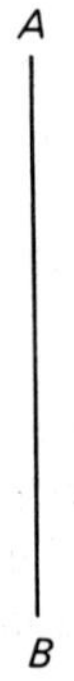

S–14–10. Construct a triangle from segments $\overline{AB}$, $\overline{CD}$, and $\overline{EF}$.

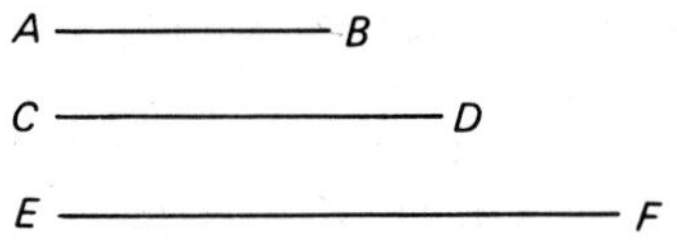

INTRODUCTION

It is necessary for tradespeople and technicians to be able to construct simple sketches of their work. Sketches help to clarify and save valuable time. The purpose of this chapter is to teach the skills necessary to construct simple geometric figures.

DEFINITIONS

A *compass* is a tool used to draw circles or portions of circles. It is also used to copy lengths of line segments. One type of compass is shown.

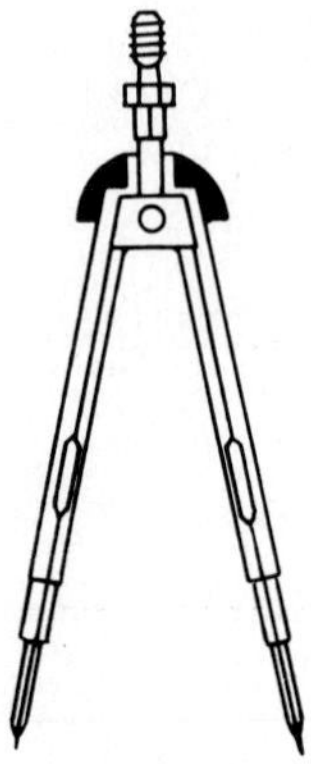

A *staightedge* is used to draw lines. A ruler is generally used as a straightedge, but anything may be used if it has a straight edge. The opening between two points of the compass is called the *radius*. The letters O and P are frequently used as point of identification in geometric construction.

BASIC CONSTRUCTIONS

Constructions are best learned through practice. The first figure constructed is a circle. Open the compass so that the metal tip and lead tip are an inch or two apart. Place the metal tip on point P at the right. Draw or "swing" a circle with the compass. Point P is the center of the circle. Practice constructing circles similar to those shown below.

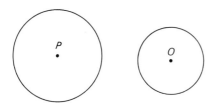

Refer to the figure and use your straightedge to draw a line from the center O to point A on the circle. This line, OA, is the radius of the circle. Drawn-in radius $\overline{OA}$ and radius $\overline{OB}$ are called *radii* of the circle. (Radii is the plural of *radius*.) Note that $\overline{OB}$ and $\overline{OC}$ form a straight line and pass through the center point O. Thus BOC is the *diameter* of the circle. The curved line AC in the figure is called an *arc*. An arc is a portion of the circumference of a circle. The symbol of the circumference for an arc is ⌢, which is placed over appropriate letters: $\overset{\frown}{AC}$. A circle may have several arcs. $\overset{\frown}{AB}$ and $\overset{\frown}{BC}$ are also arcs of circle O.

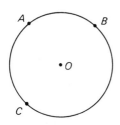

Given the radius $\overline{LM}$, construct a circle using M as the center point.

Place two additional letters, J and G, on the circumference of the circle. Identify the three arcs of circle M.

To copy a line segment, both a compass and a straightedge are used. To copy line segment $\overline{JG}$, illustrated below, draw a "working line" that is longer than the given line segment. Use a straightedge to do this.

Next, open the compass to the length of line segment $\overline{JG}$. Place the metal point on J and the lead point on G. Then place the metal point on J of the working line and "arc" the working line. At the point where the arc crosses the working line, identify point G.

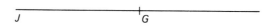

Copy the lines shown below on the right side of the page.

Example 14–1

PROBLEM: Bisect a given line segment.

SOLUTION: To bisect a line means to divide it into two equal parts. First, open the compass to a position greater than half of the length of the line segment. Place the metal tip at point K and swing an arc as shown.

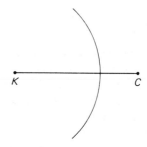

Second, keep the compass at the same opening and swing an arc from point C. Label the points where the arcs intersect as X and Y.

Chapter 14 / Construction of Simple Geometric Figures

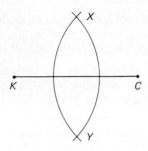

Third, sketch a line from the point of intersection X to the point of intersection Y. Where line XY intersects line $\overline{KC}$, label the point Z. Point Z is the midpoint of line KC. Thus

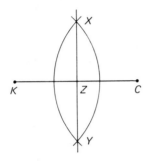

Example 14–2

PROBLEM: Bisect a given angle.

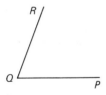

SOLUTION: To bisect an angle means to divide it into two equal angles. First, place the metal tip of your compass at the vertex of the angle (point Q) and swing an arc through the rays of the angle making point A and B. The compass opening should be $\frac{1}{2}$ to $\frac{3}{4}$ the length of $\overrightarrow{PO}$. Label points A and B.

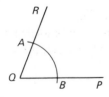

Second, swing arcs from A and B and label their point of intersection as point D.

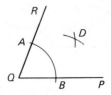

Third, draw line QD. Line QD bisects angle PQR to form two equal angles.

Angle PQD = angle RQD   or   ∠PQD = ∠RQD

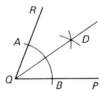

Example 14–3

PROBLEM: Copy a given angle

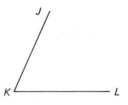

SOLUTION: To copy an angle means to produce another angle that is equal to the given angle. First, draw a working line and label it QR. This line should be somewhat longer than line KL above.

Second, swing an arc from point K above to form points N and M.

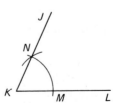

Section Three / Applied Geometry

Third, swing a similar arc on the working line from point Q. Identify the point of intersection as point X.

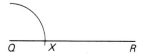

Fourth, set the compass so that the metal tip is on point M and the lead tip is on point N. Fifth, place the metal tip on point X above and swing an arc so that it intersects the other arc. Label this point of intersection as point Y.

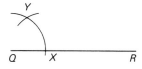

Sixth, draw a line from point Q through point Y to form line QP. Angle LKJ = angle RQP, or

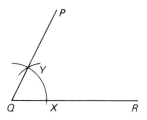

Example 14–4

PROBLEM: Divide a line into a given number of equal parts.

SOLUTION: To divide the given line $\overline{SA}$ into three equal parts, first draw a working line about 20 to 50° above line SA to form working angle MSA.

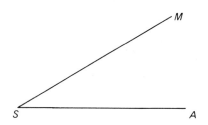

Second, divide the line SM into three equal parts by opening the compass to about $\frac{1}{3}$ of SM and swing an arc to E. From point E, swing the same arc to F. From F, swing the same arc to G. Draw a line from point G to point A.

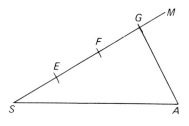

Third, using point F as a center, copy angle SGA and draw line FB.

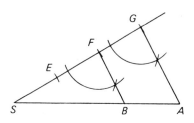

Fourth, using point E as a center, copy angle SGA and draw line EK.

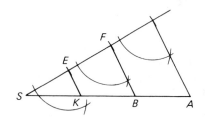

Thus $\overline{SK} = \overline{KB} = \overline{BA}$.

The same process may be used to divide any straight line into any number of equal parts.

Chapter 14 / Construction of Simple Geometric Figures

EXERCISE 14-1

Do the following constructions as indicated.

14-1. Bisect line $\overline{AB}$.

14-2. Bisect line $\overline{CD}$.

14-3. Bisect line $\overline{EF}$.

14-4. Bisect angle *ABC*.

14-5. Bisect angle *EFG*.

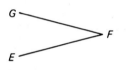

14-6. Bisect angle *HIJ*.

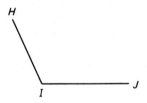

14-7. Copy angle *KLM*.

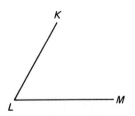

14-8. Copy angle *NOP*.

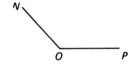

14-9. Copy angle *QRS*.

14-10. Divide line $\overline{AB}$ into three equal parts.

Section Three / Applied Geometry

14–11. Divide line $\overline{CD}$ into four equal parts.

14–12. Divide line $\overline{EF}$ into five equal parts.

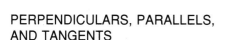

PERPENDICULARS, PARALLELS, AND TANGENTS

Walls of a building are constucted straight up from the floor so that they form a 90° angle. We say the wall is perpendicular to the floor. Thus, when one line is *perpendicular* to another, it forms two 90° angles. When two lines run side by side and never cross no matter how far they are extended, we say the lines are *parallel*. When a line touches a circle at only one point, we say that the line is *tangent* to the circle. In the next series of examples, you will learn how to construct perpendicular and parallel lines and tangents.

Example 14–5

PROBLEM: Construct a perpendicular line from a point on a line.

SOLUTION: First, place the compass point on point P, swing arcs on each side of point P, and label them X and Y.

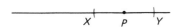

Second, open the compass and swing arcs above and below the line from points X and Y to form points A and B.

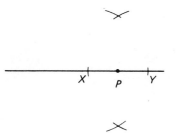

Third, draw AP using point B as a check point. $\overline{AP}$ will be perpendicular to the line at point P. Thus angles YPA and XPA are 90°, and PA is ⊥ to XY.

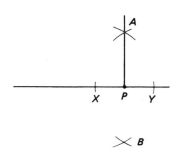

Example 14–6

PROBLEM: Construct a perpendicular to line $\overline{AB}$ from point P.

SOLUTION: First, open the compass slightly more than the distance from point P to $\overline{AB}$ and arc the line in two places, J and K.

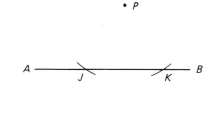

Second, using J and K as centers, swing arcs below $\overline{AB}$ and label the new point L.

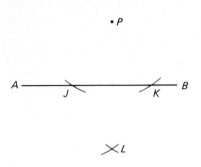

Third, draw the line segment from point P to $\overline{AB}$ using point L as the guide point to determine the line. Label the point of intersection as point Q. Thus PQ is perpendicular to $\overline{AB}$.

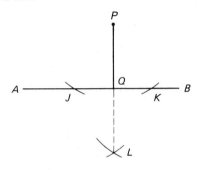

Example 14–7

PROBLEM: Construct a line parallel to a given line.

SOLUTION: First, draw working line JL, which intersects line $\overline{XY}$ at about a 45° angle, as shown.

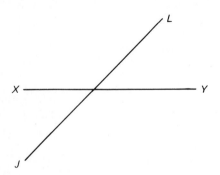

Second, swing an arc above $\overline{XY}$ at the intersection of $\overline{XY}$ and JL.

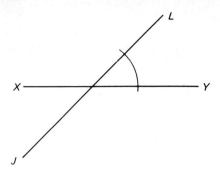

Third, swing an arc from a point on JL which might be identified as point K.

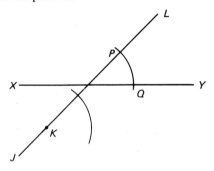

Fourth, open the compass a distance equal to PQ and swing an arc from the point where the previous arc intersected $\overline{JL}$ to intersect the previous arc at point W.

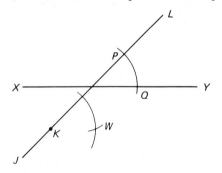

Fifth, draw line VT through points K and W. Thus $\overline{XY}$ is parallel to $\overline{VT}$, or $XY \| VT$.

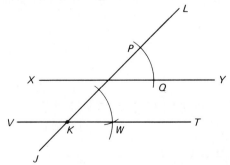

Example 14-8

PROBLEM: Construct a tangent to a circle from a given point.

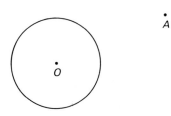

SOLUTION: First, draw a line from point A to the center of the circle, point O.

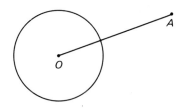

Second, construct a line that bisects $\overline{AO}$ as shown. Label the point of bisection as point K.

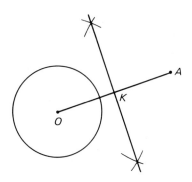

Third, swing an arc from point K to intersect point O and to create points E and D on the circle.

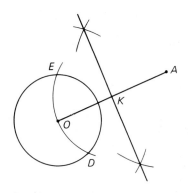

Fourth, draw rays AE and AD from point A to intersect at points E and D, respectively. Thus $\overrightarrow{AD}$ and $\overrightarrow{AE}$ are tangent to circle O.

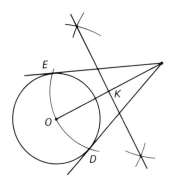

EXERCISE 14-2

Construct the following perpendiculars, parallels, and tangents as indicated.

14-13. Construct a perpendicular to the line at point X.

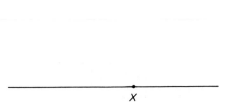

14-14. Construct a perpendicular to point P.

14–15. Construct a perpendicular to point K.

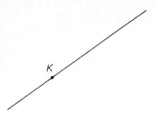

14–16. Construct a perpendicular to the line from point L.

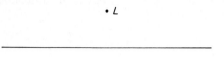

14–17. Construct a perpendicular to the line from point J.

14–18. Construct a perpendicular to the line from point K.

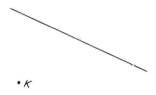

14–19. Construct a line parallel to the given line.

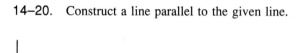

14–20. Construct a line parallel to the given line.

14–21. Construct a line parallel to the given line.

14–22. Construct tangents to circle L from point M.

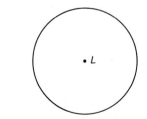

14-23. Construct tangents to circle X from point Y.

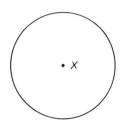

14-24. Construct tangent to circle S from point T.

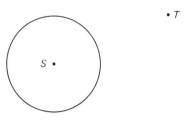

TRIANGLES, SQUARES, AND HEXAGONS

It is frequently helpful to construct simple geometric figures to understand complicated industrial problems. In this section we construct triangles, squares, and hexagons.

Example 14–9

PROBLEM: Construct a triangle from three given line segments.

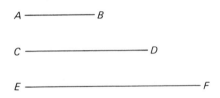

SOLUTION: First, draw a working line and label it.

Second, measure line $\overline{EF}$ with a compass and enter the points on the working line XY.

Third, measure line $\overline{AB}$ with the compass and swing an arc from point E.

Fourth, measure line $\overline{CD}$ with the compass and swing an arc from point F so that it intersects the arc from point E. Designate the point of intersection as point G.

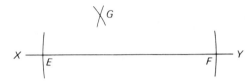

Fifth, draw in line segments $\overline{EG}$ and $\overline{FG}$.

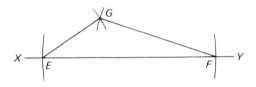

Sixth, measure with a compass to ensure that

$$\overline{AB} = \overline{EG}$$
$$\overline{CD} = \overline{FG}$$
$$\overline{EF} = \overline{EF}$$

Example 14–10

PROBLEM: Construct a square given the length of one side.

SOLUTION: First, draw a working line.

Second, measure line $\overline{KJ}$ and enter the points on the working line.

Third, construct perpendiculars at both points K and J.

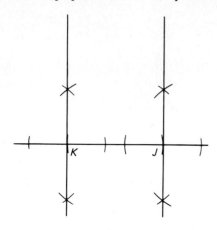

Fourth, measure the length of line segment $\overline{KJ}$ and strike an arc on the perpendiculars from points K and J. Enter points L and M.

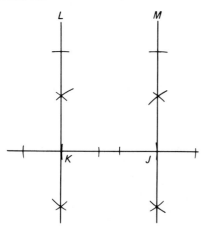

Fifth, draw in line segments $\overline{KL}$, $\overline{LM}$, and $\overline{JM}$.

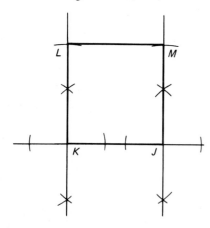

Sixth, measure with a compass to ensure that

$$\overline{KJ} = \overline{KL} = \overline{LM} = \overline{JM}$$

Thus $JKLM$ is a square.

Example 4–11

PROBLEM: Construct a regular hexagon, given the length of one side.

SOLUTION: Note that in a regular hexagon all sides are equal in length and all angles are equal. First, draw a circle using line segment $\overline{LM}$ as the radius of the cicle.

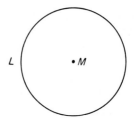

Second, using the same radius, swing an arc from point L to point A.

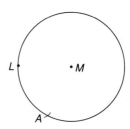

Third, using the same radius, swing an arc from point A to point B, from point B to point C to point D, and from point D to point E.

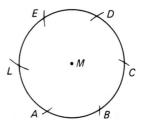

Fourth, draw in line segments $\overline{LA}$, $\overline{AB}$, $\overline{BC}$, $\overline{CD}$, $\overline{DE}$, and $\overline{EL}$.

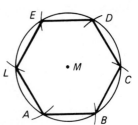

Section Three / Applied Geometry

Example 14–12

PROBLEM: Construct a 60° angle.

SOLUTION: Recall that all angles of an equilateral triangle are equal to 60°. First, draw line segment CK on working line MO.

Second, from points C and K on working line MO swing arcs of equal lengths. Label the point of intersection as point H.

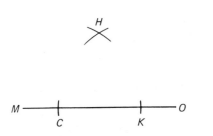

Third, draw line segments $\overline{CH}$ and $\overline{KH}$.

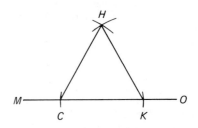

Fourth, since all sides are equal length, each interior angle is 60° and each exterior angle is 120°.

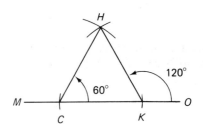

EXERCISE 14–3

Construct the following triangles, squares, rectangles, and angles as indicated.

14–25. Construct a triangle given the line segments.

A ——————— B
C ———————— D
E —————————— F

14–26. Construct a triangle given the line segments.

J ——————————— K
L ————— M
N ———————————— O

14–27. Construct a triangle given the line segments.

P ——— Q
R ———— S
T ——— U

14–28. Construct a square given line segment $\overline{AB}$.

A ——— B

Chapter 14 / Construction of Simple Geometric Figures

14–29. Construct a rectangle given the line segments CD and EF.

C ——————— D
E ——————— F

14–30. Construct a rectangle given the line segments GH and IJ.

G ————— H
I ——————— J

14–31. Construct a 120° angle.

14–32. Construct a 30° angle.

14–33. Construct a 45° angle.

14–34. Construct a 135° angle.

14–35. Construct a regular hexagon given the line segment $\overline{MN}$.

M ————————— N

14–36. Construct a parallelogram given the line segments OP and QR, with angles of 60° and 120°.

O ————————— P
Q ——————— R

THINK TIME

Corporations often create special symbols, called *logos*, to represent their companies in the minds of consumers. Study the three logos that appear below, and then design some of your own using the compass and straightedge.

PROCEDURES TO REMEMBER

1. To copy a line segment:
   (a) Draw a working line with a straightedge.
   (b) Measure the given line segment with a compass.
   (c) Indicate a point on the working line and swing an arc with the compass.
   (d) The new line segment extends from the point on the working line to the point where the arc crossed the line.

2. To bisect a line segment:
   (a) Open the compass to a position greater than half the length of the line segment.
   (b) Swing an arc from each endpoint.
   (c) Connect the points of intersection above and below the given line segment with a straightedge.
   (d) This perpendicular line will bisect the given line segment.

3. To bisect an angle:
   (a) Swing an arc from the vertex of the angle so that it intersects both rays of the angle.
   (b) Swing an arc from each of the intersection points on the rays.
   (c) From the point of intersection of these arcs, draw a line to the vertex using a straightedge.
   (d) This line bisects the angle, thus creating two equal angles.
4. To copy an angle:
   (a) Draw a working line with a straightedge.
   (b) Indicate a point on the working line to serve as a vertex.
   (c) Swing the compass on the given angle and on the working line.
   (d) Set the compass on the point of intersection of one of the two rays of the given angle and open it so that it touches the point of intersection on both rays.
   (e) Set the compass on the point of intersection on the working line. Swing the same arc as measured in step (d).
   (f) Draw a line from the vertex of the working line through the intersecting arcs above the working line.
   (g) This line will be the second ray of the new angle.
5. To divide a line into a given number of equal segments:
   (a) Draw a working line 20° to 50° above the given line segment so that the two lines form an angle with a common vertex.
   (b) Divide the working line into the given number of segments by swinging preset arcs from the vertex to the first intersection of the working line; then from the first intersection, swing the second preset arc; from the second intersection, swing the third preset arc; and so on.
   (c) Draw a line from the last intersection to the endpoint of the given line segment.
   (d) Copy each angle and draw lines to intersect the given line segment.
   (e) Each intersection will create equal line segments.
6. To construct a perpendicular line from a point on a line:
   (a) Place the compass on the given point on the line segment and swing an arc to cross the line segment at two different points.
   (b) From these two intersections, swing two arcs from each intersection one below and one above the given line segment.
   (c) Draw a straight line from these points of intersection to intersect the given line in a perpendicular fashion.
7. To construct a perpendicular from a given point:
   (a) Open the compass and swing an arc from the given point so that two intersections are made on the given segment.
   (b) From these points of intersection, swing two additional arcs so that they intersect on the opposite side of the line from the given point.
   (c) Connect the given point with the point of intersection of two arcs. This line will be perpendicular to the given line segment.
8. To construct a line parallel to a given line:
   (a) Construct a working diagonal intersecting the given line segment at about 45°.
   (b) Swing an arc from the point of intersection of the given line and the working line.
   (c) Swing an arc from another point on the working line to intersect the first one.
   (d) Draw a line from the point on the working line through the point of intersect of the two arcs.
   (e) The line will be parallel to the given line.
9. To construct a tangent to a circle from a given point:
   (a) Draw a line from the given point to the center of the circle.
   (b) Construct a bisector of the line drawn from the given point to the center of the circle.
   (c) Swing an arc from the midpoint of the line to intersect the center of the circle and the circumference of the circle at two points.
   (d) Draw lines from the given point to the two points on the circumference of the circle.
   (e) These two lines are tangents to the circle.
10. To construct a triangle from three given lines:
    (a) Draw a working line. Measure one of the given line segments with a compass, and enter this on the working line.
    (b) Measure a second line segment and swing an arc from one endpoint of the given line segment on the working line.
    (c) Measure the third line segment and swing an arc from the other endpoint of the given line segment so that the arc intersects the arc swing in step (b).
    (d) Draw lines from the point of intersection to each end of the given line segment.
11. To construct a square, given one side:
    (a) Draw a working line, measure the given side, and enter this on the working line.
    (b) Construct perpendiculars to the endpoints of the given side on the working line.

Chapter 14 / Construction of Simple Geometric Figures

(c) Measure the length of the given side, and from each of the endpoints of the given side, swing an arc on the perpendicular lines.
(d) Draw a line between the perpendiculars to form a square.
(e) Draw a line between the perpendiculars to form a square.

12. To construct a regular hexagon, given the length of one side:
   (a) Draw a circle using the line segment as the radius of the circle.
   (b) Using the same radius, swing an arc from the point where the radius touches the circle to cut the edge of the circle on both sides of the point.
   (c) From the point where the previous arc intersected the circle, swing an additional arc. Continue to do this until the edge of the circle has been cut by arcs into six equal lengths.
   (d) Draw line segments to connect adjacent intersections.

13. To construct a 60° angle:
   (a) Draw a line segment.
   (b) From one end of the line segment, swing an arc of equal length to the line segment.
   (c) From the other end of the line segment, swing an arc of equal length to the line segment.
   (d) At the point of intersection, draw line segments to the endpoints of the original line segments.
   (e) The inside angles of an equilateral triangle are 60° angles.

CHAPTER SUMMARY

1. The following definitions are necessary for the construction of simple geometric forms.

   (a) A *compass* is a tool used to draw circles or portions of circles and to measure line segments.
   (b) A *straightedge* is used to draw straight lines.
   (c) A *radius* is the opening between two points on a compass.
   (d) *Radii* is the plural of radius.
   (e) An *arc* is a portion of the circumference of a circle.
   (f) *Perpendicular lines* form angles of 90°.
   (g) Two lines are *parallel* when they run side by side and never intersect.
   (h) A *tangent* line touches a circle at only one point.

2. Procedures to remember, included in this chapter, are:
   (a) Copy a line segment.
   (b) Bisect a line segment.
   (c) Bisect an angle.
   (d) Copy an angle.
   (e) Divide a line segment into a given number of equal segments.
   (f) Construct a perpendicular line from a point on a line.
   (g) Construct a perpendicular from a given point.
   (h) Construct a line parallel to a given line.
   (i) Construct a tangent to a circle from a given point.
   (j) Construct a triangle from three given line segments.
   (k) Construct a square given one side.
   (l) Construct a regular hexagon given the length of one side.
   (m) Construct a 60° angle.

CHAPTER TEST

T–14–1. Copy line segment $\overline{ST}$.

T–14–2. Copy angle XYZ.

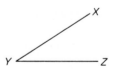

T–14–3. Bisect $\overline{PQ}$.

T–14–4. Bisect angle KLM.

P ———————— Q

T-14-5. Divide $\overline{FG}$ into four equal parts.

F ———————————— G

T-14-6. Construct a perpendicular on $\overline{AB}$ at X.

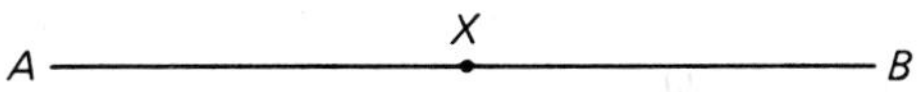

A ————————•———————— B
               X

T-14-7. Construct a line parallel to $\overline{DE}$.

D ———————————— E

T-14-8. Construct a perpendicular to $\overline{AB}$ from point Z.

A ———————————— B

• Z

T-14-9. Using $\overline{JK}$, construct a hexagon with each side equal in length to $\overline{JK}$.

J ——— K

T-14-10. Construct a 60° angle.

T-14-11. Construct a tangent to circle O from point Y.

T-14-12. Construct a triangle from the segments shown below.

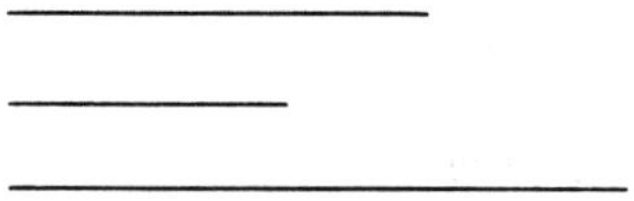

———————
————
————————

T-14-13. Construct a square given side $\overline{LM}$.

L ——— M

T-14-14. Construct a perpendicular to $\overline{WX}$.

W ———————————— X

T-14-15. Divide $\overline{QR}$ into six equal parts.

Q ———————————— R

Chapter 14 / Construction of Simple Geometric Figures

section four: trigonometry

15

FUNDAMENTALS OF TRIGONOMETRY

OBJECTIVES

1. To identify and understand the parts of triangles.
2. To identify and understand the properties of right triangles.
3. To identify and understand the trigonometric ratios of the sine, cosine, tangent, cotangent, secant, and cosecant.
4. To solve similar triangles using ratio and proportion techniques.

SELF-TEST

This test will measure your understanding of basic trigonometry. Successful completion will indicate that you are ready to do more complex trigonometry problems in Chapters 16 and 17.

S-15-1. How many degrees are in ∠A in Figure 15-1?

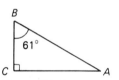

Figure 15-1

S-15-2. How many degrees are in ∠E in triangle DEF in Figure 15-2?

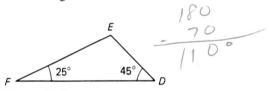

Figure 15-2

S-15-3. Letter the sides of triangle GHI in Figure 15-3.

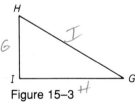

Figure 15-3

S-15-4. Letter the sides of triangle MLK in Figure 15-4.

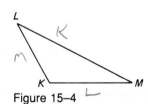

Figure 15-4

241

S-15-5. Letter the angles of the triangle in Figure 15-5.

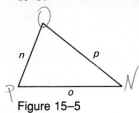

Figure 15-5

S-15-6. Given similar triangles QRS and TUV, find q in Figure 15-6.

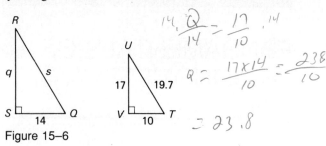

Figure 15-6

$$\frac{q}{14} = \frac{17}{10}$$

$$q = \frac{17 \times 14}{10} = \frac{238}{10}$$

$$= 23.8$$

S-15-7. Find s in Figure 15-6.

$$\frac{s}{14} = \frac{19.7}{10} \Rightarrow s = \frac{19.7 \times 14}{10} = \frac{275.8}{10}$$

$$s = 27.58$$

S-15-8. Given similar triangles ABC and DEF in Figure 15-7, find ∠A.

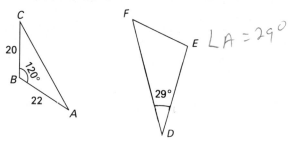

∠A = 29°

Figure 15-7

S-15-9. Find the size of ∠F in Figure 15-7.

∠F = 31°

S-15-10. Given similar triangles GHI and JKL in Figure 15-8, find the length of i to the nearest tenth.

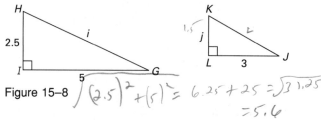

Figure 15-8

$\sqrt{(2.5)^2 + (5)^2} = \sqrt{6.25 + 25} = \sqrt{31.25}$

$= 5.6$

S-15-11. Find the length of j to the nearest tenth in Figure 15-8.

$$\frac{j}{3} = \frac{2.5}{5} \Rightarrow j = \frac{2.5 \times 3}{5} \Rightarrow j = 1.5$$

S-15-12. Find the length of l to the nearest tenth in Figure 15-8.

$\sqrt{(1.5)^2 + (3)^2} \sqrt{2.25 + 9} \sqrt{11.25}$

$= 3.35$

S-15-13. The sine of angle N is defined as the ratio $\frac{n}{o}$. What is the sine of angle M in Figure 15-9?

$\frac{o}{\sqrt{}} \frac{m}{o}$

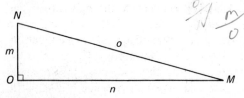

Figure 15-9

S-15-14. What is the tangent ratio of angle N in Figure 15-9?

$TAN = \frac{n}{m}$

S–15–15. What is the cosine ratio of angle *M* in Figure 15–9?

S–15–16. What trigonometric function equals the ratio $\frac{q}{r}$ in Figure 15–10?

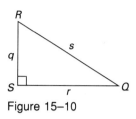
Figure 15–10

S–15–17. In Figure 15–10, what trigonometric function equals the ratio $\frac{r}{q}$?

S–15–18. In Figure 15–10, what two trigonometric functions equal the ratio $\frac{r}{s}$?

S–15–19. If the trigonometric function of the sine of angle *A* is 0.5000, what is the size of angle *A*?

S–15–20. If the trigonometric function of the tangent of angle *R* is 0.577, what is the size of angle *R*?

INTRODUCTION

Trigonometry is a combination of arithmetic, algebra, and geometry that is used by tradespeople and technicians to solve applied problems of industry. The word *trigonometry* is derived from the Greek words, *trigonon* and *metria*, which mean "triangle measure." However, trigonometry has expanded and includes the solution of problems dealing with squares, circles, rectangles, and other geometric figures. Trigonometry, or "trig" as it is commonly called, is essential for solving certain problems—solutions that could not be found by any other branch of mathematics.

DEFINITIONS

All triangles have six parts: three sides and three angles. Finding the six parts of a triangle is called *solving the triangle*. A *right triangle* is a triangle that has a right or 90° angle.

Figure 15–11 is a right triangle. Study the names of

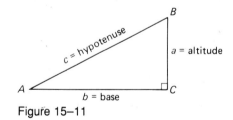
Figure 15–11

the parts of the triangle. Angle *C* is a right angle. The letter *C* is generally assigned to the right angle. The symbol ∟ is used to identify a right angle. Angles are identified by capital letters. Each side is identified by a lowercase letter, the same letter as that given to the angle opposite the side. Thus side *a* is opposite angle *A*, side *b* is opposite angle *B*, and side *c* is opposite angle *C*. Each of the sides has a name. Side *a* is the altitude, side *b* is the base, and side *c* is the hypotenuse. However, the positioning and the shape of the right triangle may not always identify a base or altitude. The side opposite the right angle is always called the *hypotenuse*.

PROPERTIES OF RIGHT TRIANGLES

The problems and solutions that follow will illustrate other properties of right triangles.

Example 15–1

PROBLEM: Find the sum in degrees of the angles of the right triangle shown in Figure 15–12.

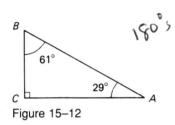

Figure 15–12

SOLUTION: The sum, in degrees, of the angles of any triangle will always be 180°. Measure the two unknown angles with a protractor given that ∠C equals 90°. If only one angle is unknown, subtract the sum of the two known angles from 180°. Then check by adding the degrees of the angles to equal 180°. Thus 90° + 61° + 29° = 180°.

EXERCISE 15–1

Name the triangles and their parts as indicated in the following problems.

15–1. What type of triangle is *ABC* in Figure 15–15?

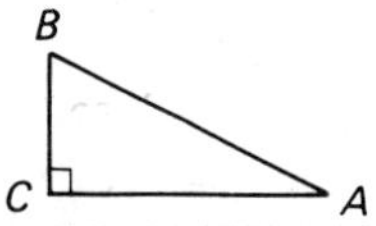

Right Triangle

Figure 15–15

15–3. Letter the sides of the triangle in Figure 15–16.

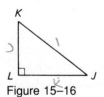

Figure 15–16

Another way to prove the total degrees of a triangle is 180° is to tear the angles from a paper triangle and arrange them together to form a straight angle or 180°. See Figure 15–13.

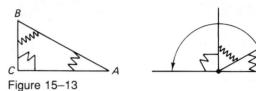

Figure 15–13

Example 15–2

PROBLEM: Find the total degrees of the angles of *any* triangle as shown by Figure 15–14.

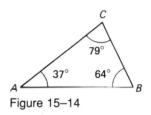

Figure 15–14

SOLUTION: Measure two unknown angles with a protractor if necessary. If one angle is known, measure only one angle. Second, subtract the sum of the degrees of the two known angles from 180°. Third, check by adding the degrees of the angles to total 180°. Thus 37° + 64° + 79° = 180°.

15–2. What type of angle is ∠C?

90° Right

15–4. Letter the sides of the triangle in Figure 15–17.

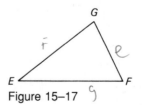

Figure 15–17

244   Section Four / Trigonometry

15–5. Letter the angles of the triangle in Figure 15–18.

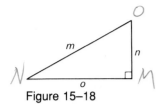
Figure 15–18

15–6. Label the sides of the triangle in Figure 15–19.

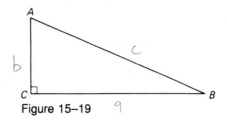
Figure 15–19

15–7. Label the sides of the triangle in Figure 15–20.

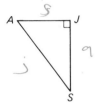
Figure 15–20

15–8. Determine the size of the missing angle in Figure 15–21.

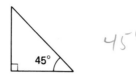

45°

Figure 15–21

15–9. Determine the size of the missing angle in Figure 15–22.

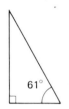

$180 - 151 = 29°$

Figure 15–22

15–10. Determine the size of the missing angle in Figure 15–23.

$180 - 84 - 64 = 32°$

Figure 15–23

SIMILAR TRIANGLES

Triangles are *similar* if their corresponding angles are equal. The triangles in Figure 15–24 are similar right triangles. The equality of angles can be shown by comparing the ratios of known sides. Study the triangles and the ratios given below.

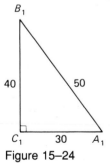
Figure 15–24

$$\frac{C_1A_1}{A_1B_1} = \frac{C_2A_2}{A_2B_2} = \frac{C_3A_3}{A_3B_3}$$

$$\frac{30}{50} = \frac{22.5}{37.5} = \frac{15}{25}$$

$$0.6 = 0.6 = 0.6$$

Pick any 2 sides for the ratio

Equal ratios of the lengths of the sides indicate equality of angles. Study the triangles shown in Figure 15–25, and the following ratios, which prove that the corresponding angles are equal.

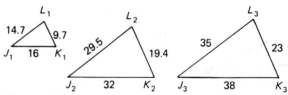
Figure 15–25

$$\frac{J_1L_1}{J_1K_1} = \frac{J_2L_2}{J_2K_2} = \frac{J_3L_3}{J_3K_3}$$

$$\frac{14.7}{16} = \frac{29.5}{32} = \frac{35}{38}$$

$$0.92 = 0.92 = 0.92$$

The fact that the ratios of the sides of similar triangles are equal is a basic principle of trigonometry. This principle is the foundation of finding solutions to trigonometry problems.

Chapter 15 / Fundamentals of Trigonometry

Example 15-3

PROBLEM: Given two similar triangles (see Figure 15-26), determine the lengths of the remaining sides.

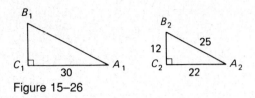
Figure 15-26

SOLUTION: Corresponding sides of similar triangles have similar ratios, thus ratios can be set up and unknowns can be found using simple algebra. Set up the ratios.

$$\frac{C_1A_1}{C_2A_2} = \frac{A_1B_1}{A_2B_2} = \frac{B_1C_1}{B_2C_2}$$

Substitute known values into the proper proportion:

$$C_1A_1 = 30 \quad C_2A_2 = 22$$
$$A_2B_2 = 25 \quad B_2C_2 = 12$$

$$\frac{C_1A_1}{C_2A_2} = \frac{A_1B_1}{A_1B_2}$$

$$\frac{30}{22} = \frac{A_1B_1}{25}$$

Solve for the unknown A_1B_1.

$$\frac{30 \times 25}{22} = A_1B_1$$
$$34.1 = A_1B_1$$

Then substitute known values.

$$C_1A_1 = 30 \quad C_2A_2 = 22 \quad B_2C_2 = 12$$

$$\frac{C_1A_1}{C_2A_2} = \frac{B_1C_1}{B_1C_2}$$

$$\frac{30}{22} = \frac{B_1C_1}{12}$$

Solve for the unknown B_1C_1.

$$\frac{30 \times 12}{22} = B_1C_1$$

$$B_1C_1 = 16.\overline{36}$$
$$B_1C_1 = 16.4$$

EXERCISE 15-2

Solve the following similar triangles.

15-11. Given the similar triangles $A_1B_1C_1$ and $A_2B_2C_2$ in Figure 15-27, solve for A_2B_2.

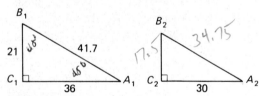
Figure 15-27

15-13. Given the similar triangles $J_1K_1L_1$ and $J_2K_2L_2$ in Figure 15-28, solve for J_1K_1.

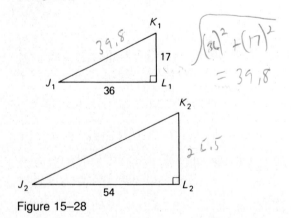
Figure 15-28

15-12. Solve for B_2C_2 in Figure 15-27.

$B_2C_2 = 17.5$

15-14. Solve for L_2K_2 in Figure 15-28.

$\frac{54}{36} = \frac{x}{17}$

$\frac{17}{36} = \frac{x}{54} \quad x = \frac{17 \times 54}{36} = 25.5$

15–15. Solve for K_2J_2 in Figure 15–28.

$\sqrt{(25.5)^2 + (54)^2} = 59.7$

$\frac{46}{92} = \frac{y}{46} = y = \frac{46 \times 46}{92} = 23$

15–16. Given the similar triangles DEF and XYZ in Figure 15–29, solve for y.

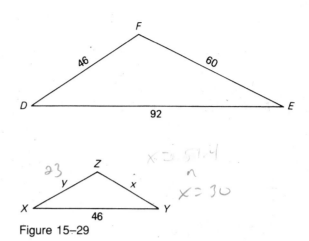

$x = 51.4$
$x = 30$

Figure 15–29

15–17. Solve for x in Figure 15–29.

$\sqrt{(23)^2 + (46)^2} = 51.4$

$\frac{60}{92} = \frac{x}{46} = x = \frac{60 \times 46}{92}$

$x = 30$

15–18. Given the similar triangles MNO and EFG in Figure 15–30, solve for e.

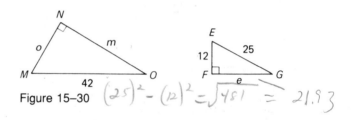

Figure 15–30 $\sqrt{(25)^2 - (12)^2} = \sqrt{481} = 21.93$

15–19. Solve for o in Figure 15–30.

$\frac{12}{21.93} = \frac{o}{42} = \frac{12 \times 42}{21.93} = o = 22$

$\frac{o}{12} = \frac{42}{25} = o = 20.16$

15–20. Solve for m in Figure 15–30.

$\frac{42}{25} = \frac{m}{21.93} = m = 36.84$

TRIGONOMETRIC RATIOS

The sides of right triangles are named in relation to their angles. In triangle ABC, Figure 15–31, angle A indicates that side b is the adjacent side, side a is the opposite side, and c is the hypotenuse.

Regarding angle B, side b is the opposite side, side a is the adjacent side, and side c is the hypotenuse. A rule to follow for naming the sides of a right triangle concerning particular angles is: (1) the side opposite the right angle is called the *hypotenuse*; (2) the side that is a part of the angle but is not the hypotenuse is called the *adjacent side*; and (3) the side opposite or farthest from the angle is called the *opposite side*.

The ratios of the sides of right triangles are identified by specific names. These ratios and their identification are necessary in solving trig problems. In triangle ABC, angle A determines the names of the sides of the triangle.

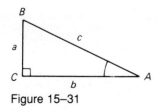

Figure 15–31

Chapter 15 / Fundamentals of Trigonometry

See Figure 15-32. The names of the trig ratios concerning angle A are as follows:

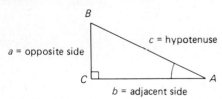

Figure 15-32

$$\text{sine of angle } A = \frac{\text{opposite side}}{\text{hypotenuse}} \text{ or}$$

$$\sin A = \frac{\text{opp}}{\text{hyp}} = \frac{a}{c}$$

$$\text{cosine of angle } A = \frac{\text{adjacent side}}{\text{hypotenuse}} \text{ or}$$

$$\cos A = \frac{\text{adj}}{\text{hyp}} = \frac{b}{c}$$

$$\text{tangent of angle } A = \frac{\text{opposite side}}{\text{adjacent}} \text{ or}$$

$$\tan A = \frac{\text{opp}}{\text{adj}} = \frac{a}{b}$$

There are three additional trigonometric functions that are reciprocals of the sin, cos, and tan functions. They are defined as follows in relation to angle A in Figure 15-32:

$$\text{cotangent of angle } A = \cot A = \frac{\text{adj}}{\text{opp}}$$

$$\cot A = \frac{b}{a} \quad \text{which is the reciprocal of}$$

$$\tan A = \frac{a}{b}$$

$$\text{secant of angle } A = \sec A = \frac{\text{hyp}}{\text{adj}}$$

$$\sec A = \frac{c}{b} \quad \text{which is the reciprocal of}$$

$$\cos A = \frac{b}{c}$$

$$\text{cosecant of angle } A = \csc A = \frac{\text{hyp}}{\text{opp}}$$

$$\csc A = \frac{c}{a} \quad \text{which is the reciprocal of}$$

$$\sin A = \frac{a}{c}$$

It is necessary to know the trig ratios to solve trig problems. You should know the trig ratios as well as the multiplication facts. The ratios of the sides of right triangles are also known as the *trigonometric functions*. Since the angles will not always be designated as A, B, and C, it is important to identify the trigonometric functions with the names of the sides and not with given letters. The two examples illustrate how to define the three basic trigonometric functions.

Example 15-4

PROBLEM: Identify the three basic trigonometric functions of angle D and angle E in triangle DEF in Figure 15-33.

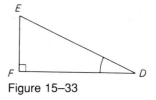

Figure 15-33

SOLUTION: Name the sides in relation to angle D in Figure 15-34. Identify the trigonometric functions.

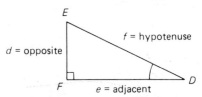

Figure 15-34

$$\sin D = \frac{\text{opp}}{\text{hyp}} = \frac{d}{f}$$

$$\cos D = \frac{\text{adj}}{\text{hyp}} = \frac{e}{f}$$

$$\tan D = \frac{\text{opp}}{\text{adj}} = \frac{d}{e}$$

Then identify the sides in relation to angle E as shown in Figure 15-35. Finally, identify the trigonometric functions.

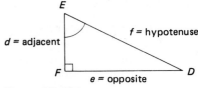

Figure 15-35

$$\sin E = \frac{\text{opp}}{\text{hyp}} = \frac{e}{f}$$

$$\cos E = \frac{\text{adj}}{\text{hyp}} = \frac{d}{f}$$

$$\tan E = \frac{\text{opp}}{\text{adj}} = \frac{e}{d}$$

Example 15-5

PROBLEM: Identify the basic trigonometric functions of angle J and angle K in triangle JKL in Figure 15-36.

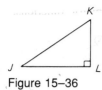
Figure 15-36

SOLUTION: Name the sides in relation to angle J as shown in Figure 15-37. Identify the trigonometric functions.

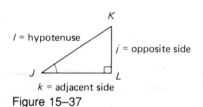
Figure 15-37

$$\sin J = \frac{\text{opp}}{\text{hyp}} = \frac{j}{l}$$
$$\cos J = \frac{\text{adj}}{\text{hyp}} = \frac{k}{l}$$
$$\tan J = \frac{\text{opp}}{\text{adj}} = \frac{j}{k}$$

Then identify the sides in relation to angle K as shown in Figure 15-38, and identify the trigonometric functions.

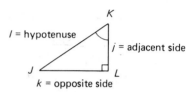
Figure 15-38

$$\sin K = \frac{\text{opp}}{\text{hyp}} = \frac{k}{l}$$
$$\cos K = \frac{\text{adj}}{\text{hyp}} = \frac{j}{l}$$
$$\tan K = \frac{\text{opp}}{\text{adj}} = \frac{k}{j}$$

EXERCISE 15-3

Identify the trigonometric functions as indicated in the following problems.

15-21. Given the right triangle HIJ in Figure 15-39, what is the sine of ∠H?

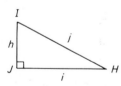
Figure 15-39

$\frac{H}{O} = \frac{h}{j}$

15-22. What is the tangent of ∠I, Figure 15-39?

$\frac{O}{A} = \frac{i}{h}$

15-23. What is the cosine of ∠H, Figure 15-39?

$\frac{A}{H} \quad \frac{i}{j}$

15-24. What is the sine of ∠I, Figure 15-39?

$\frac{O}{H} = \frac{i}{j}$

15-25. What is the cosine of ∠I, Figure 15-39?

$\frac{A}{H} = \frac{h}{j}$

15-26. What is the tangent of ∠H, Figure 15-39?

$\frac{O}{A} = \frac{h}{i}$

15–27. Letter the sides of triangle XYZ shown in Figure 15–40.

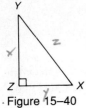
Figure 15–40

15–28. What is sin X in Figure 15–40?

$\frac{O}{H} = \frac{x}{z}$

15–29. What is tan X in Figure 15–40?

$\frac{O}{A} = \frac{x}{y}$

15–30. What is cos Y in Figure 15–40?

$\frac{A}{H} = \frac{x}{z}$

15–31. What is tan Y in Figure 15–40?

$\frac{O}{A} = \frac{y}{x}$

15–32. What is sin Y in Figure 15–40?

$\frac{O}{H} \quad \frac{y}{z}$

15–33. What is cos X in Figure 15–40?

$\frac{A}{H} \quad \frac{y}{z}$

15–34. Letter the angles of the triangle in Figure 15–41.

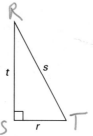
Figure 15–41

15–35. What is sin R in Figure 15–41?

$\frac{O}{H} \quad \frac{r}{s}$

15–36. What is cos T in Figure 15–41?

$\frac{A}{H} \quad \frac{r}{s}$

15–37. What is tan T in Figure 15–41?

$\frac{O}{A} = \frac{T}{R}$

15–38. Complete the lettering of the triangle in Figure 15–42.

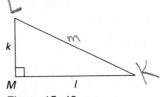
Figure 15–42

Determine the cosine of 45°.

$$\tan 45° = \frac{\text{opp}}{\text{adj}} = \frac{1}{\sqrt{2}} = \frac{1}{\sqrt{2}}\left(\frac{\sqrt{2}}{\sqrt{2}}\right)$$

$$= \frac{\sqrt{2}}{2} = \frac{1.414}{2} = 0.707$$

Thus the values of the basic trigonometric functions of a 45° angle are

$$\sin 45° = 0.707$$
$$\cos 45° = 0.707$$
$$\tan 45° = 1.000$$

EXERCISE 15–4

Determine the values of the trig functions as indicated in the following problems.

15–48. Given the triangle ABC, Figure 15–48, find b.

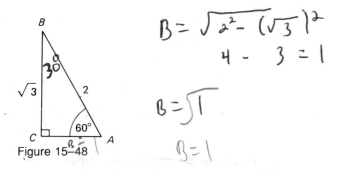

Figure 15–48

$B = \sqrt{2^2 - (\sqrt{3})^2}$
$4 - 3 = 1$
$B = \sqrt{1}$
$B = 1$

15–49. Find ∠B, Figure 15–48.

$90 - 60 = 30°$

15–50. Find the sine of 60° in Figure 15–48.

$\frac{O}{H} \quad \frac{\sqrt{3}}{2} = .8660$

15–51. Find the cosine of 60° in Figure 15–48.

$\frac{A}{H} = \frac{1}{2} = \cos = .5$

15–52. Find the tangent of 60° in Figure 15–48.

$\frac{O}{A} \quad \frac{\sqrt{3}}{1} = \sqrt{3}$

15–53. Given the right triangle JKL, Figure 15–49, determine the value of l.

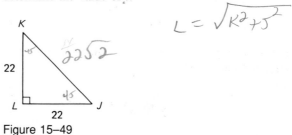

Figure 15–49

$L = \sqrt{K^2 + J^2}$
$22\sqrt{2}$

15–54. Determine the tangent of ∠J, in Figure 15–49.

$\frac{O}{A} \quad \frac{22}{22} = 1$

15–55. Determine the sine of ∠K in Figure 15–49.

$\frac{O}{H} = \frac{22}{22\sqrt{2}} = \frac{\sqrt{2}}{2} = .707106$

.781

Chapter 15 / Fundamentals of Trigonometry

15–56. Determine the cosine of ∠J in Figure 15–49.

15–57. How many degrees are there in angle J in Figure 15–49?

15–58. Angle K equals how many degrees in Figure 15–49?

15–59. Given right triangle PQR, Figure 15–50, determine the value of side p.

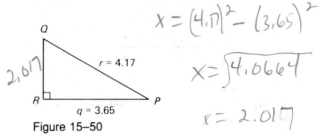

Figure 15–50

$x = (4.17)^2 - (3.65)^2$

$x = \sqrt{4.0664}$

$x = 2.017$

15–60. Determine the value of the tangent of ∠P in Figure 15–50.

$\frac{O}{A} = \frac{2.017}{3.65}$

TAN ∠P = 0.553

15–61. Determine the value of sine of ∠Q in Figure 15–50.

$\frac{O}{H} = \frac{3.65}{4.17} = .87529976$

15–62. Determine the value of the tangent of ∠Q in Figure 15–50.

TAN ∠Q $\frac{O}{A}$ $\frac{3.65}{2.017} = 1.810$

15–63. Determine the value of the cosine of ∠Q in Figure 15–50.

COS ∠Q = $\frac{A}{H}$ $\frac{2.017}{4.17} = .484$

15–64. How many degrees are there in ∠Q in Figure 15–50?

61° 04' 23"
49"

15–65. How many degrees are there in ∠P in Figure 15–50?

∠P = 29°

THINK TIME

Many students find it helpful to have a word to help them remember several things. Three acronyms, *soph*, *cash*, and *topa*, may help you remember the trig functions. The meanings of the acronyms are as follows:

soph (*s*ine = *op*posite over *h*ypotenuse)

cash (*c*osine = *a*djacent over *h*ypotenuse)

topa (*t*angent = *o*pposite over *a*djacent)

PROCEDURES TO REMEMBER

1. To find the unknown sides of similar triangles:
   (a) Set up ratios between similar sides.
   (b) Substitute the given values for the sides.
   (c) Solve for the unknown sides.

2. The formulas for basic trigonometric functions are:

$$\sin A = \frac{\text{opposite side}}{\text{hypotenuse}}$$
$$\cos A = \frac{\text{adjacent side}}{\text{hypotenuse}}$$
$$\tan A = \frac{\text{opposite side}}{\text{adjacent side}}$$
$$\cot A = \frac{\text{adjacent side}}{\text{opposite side}}$$
$$\sec A = \frac{\text{hypotenuse}}{\text{adjacent side}}$$
$$\csc A = \frac{\text{hypotenuse}}{\text{opposite side}}$$

CHAPTER SUMMARY

The following definitions are important when solving problems with trigonometric functions.

1. All triangles have six parts: three sides and three angles.
2. Finding the six parts of a triangle is called solving the triangle.
3. A right triangle is a triangle that has a right (90°) angle.
4. Capital letters are used to identify angles.
5. Lowercase letters are used to identify sides.
6. In a right triangle: the side opposite the right angle is the hypotenuse; the side opposite or farthest from a given angle is the opposite side; and the side next to a given angle is the adjacent side.
7. The sum of the degrees of the interior angles of a triangle is 180°.
8. Triangles are similar if their corresponding angles are equal.
9. Identical ratios of the lengths of the side of triangles indicate that the angles are equal.
10. The ratios of the sides of right triangles are called the trigonometric functions.
11. The value of the side opposite a 30° angle of a right triangle is one half the hypotenuse.
12. In a right triangle, the square of the hypotenuse is equal to the sum of the squares of the other two sides.

CHAPTER TEST

T–15–1. How many degrees are in angle B, triangle ABC, in Figure 15–51?

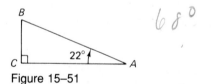

Figure 15–51

T–15–2. Find the size of angle E in triangle DEF in Figure 15–52.

Figure 15–52

T–15–3. Letter the sides of triangle GHI in Figure 15–53.

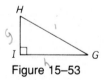

Figure 15–53

T–15–4. Letter the sides of triangle XYZ in Figure 15–54.

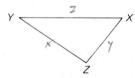

Figure 15–54

T–15–5. Letter the angles of the triangle in Figure 15–55.

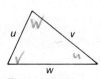

Figure 15–55

T–15–6. Given similar triangles JKL and MNO, find the length of m in triangle MNO in Figure 15–56.

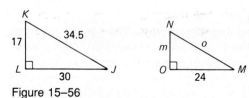

Figure 15–56

Chapter 15 / Fundamentals of Trigonometry

T–15–7. Find the length of *o* in triangle *MNO* in Figure 15–56.

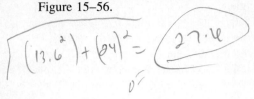

T–15–8. Given similar triangles *PQR* and *STU*, find the length of *r* in triangle *PQR* in Figure 15–57.

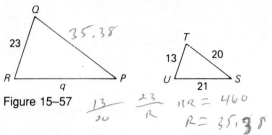

Figure 15–57

$\frac{13}{20} \quad \frac{23}{R} \quad 13R = 460$

$R = 35.38$

T–15–9. Find the length of *q* in triangle *PQR* in Figure 15–57.

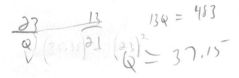

$13Q = 483$

$Q = 37.15$

T–15–10. Given similar triangles *XYZ* and *ABC*, determine the length of *z* in triangle *XYZ* to the nearest tenth in Figure 15–58.

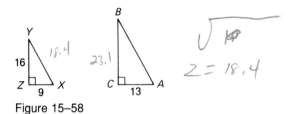

$Z = 18.4$

Figure 15–58

T–15–11. Find the length of *a* in triangle *ABC* to the nearest tenth in Figure 15–58.

$\frac{16}{9} = \frac{A}{13} \quad 9A = 208$

$A = 23.1$

T–15–12. Find the length of *c* in triangle *ABC* to the nearest tenth in Figure 15–58.

$\sqrt{702.41}$

$c = 26.15$

T–15–13. Given the right triangle *DEF*, what is the tangent ratio of angle *E* in Figure 15–59?

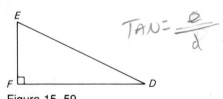

$TAN = \frac{e}{d}$

Figure 15–59

T–15–14. What is the cosine ratio of angle *D* in Figure 15–59?

$cos D = \frac{e}{f}$

T–15–15. What is the sine ratio of angle *E* in Figure 15–59?

$sine E = \frac{e}{f}$

T–15–16. In Figure 15–60, given the right triangle *GHI*, what trigonometric functions equal the ratio $\frac{h}{g}$?

cosiot H

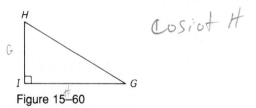

Figure 15–60

T–15–17. In Figure 15–60, what trigonometric functions equal the ratio $\frac{h}{i}$?

T–15–18. In Figure 15–60, what trigonometric functions equal ratio $\frac{g}{i}$?

T–15–19. If the tangent of angle D is 1.000, what is the size of angle D?

89.94°

T–15–20. If the sine of angle K is 0.866, what is the size of angle K?

16

SOLUTION OF RIGHT TRIANGLES

OBJECTIVES

1. To read and understand a trigonometric functions table.
2. To determine the value of trigonometric functions using a scientific calculator.
3. To solve right-triangle problems using trigonometric functions.
4. To solve applied problems using applied trigonometric functions.

SELF-TEST

This test will measure your ability to use trigonometric functions to solve right triangles. You will need a table of trigonometric functions or scientific calculator to solve these problems. Satisfactory completion will indicate your mastery of these skills and your readiness to solve oblique triangles.

S–16–1. Find the sine of 43°.

.68199836
.682

S–16–2. Find the cosine of 5°40′.

.9955

S–16–3. Find the tangent of 48°20′.

1.118

S–16–4. Find the angle whose tangent is 0.2186.

12°33′

S–16–5. Find the angle whose tangent is 4.1653.

TAN ∠A = 76.5

S–16–6. Find the tangent of 11°37′.

TAN ∠A = .2056

S–16–7. Find the angle whose cosine is 0.5476.

cos A = .5476

∠A = 56.8

S–16–8. Determine the degrees in angle B of triangle ABC in Figure 16–1.

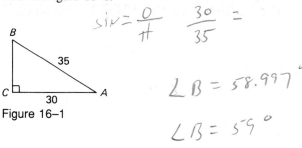
Figure 16–1

sin = O/H   30/35 =

∠B = 58.997°

∠B = 59°

S–16–9. Given triangle PQR in Figure 16–2, determine the length of q.

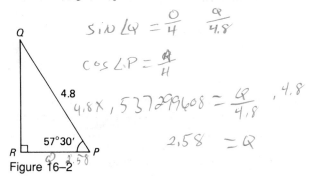
Figure 16–2

sin ∠Q = O/H   Q/4.8

cos ∠P = Q/H

4.8 × .53729608 = Q/4.8 × 4.8

2.58 = Q

S–16–10. Find the length of p in triangle PQR in Figure 16–2.

sin ∠P = O/H

4.8 × .84339/446 = P/4.8 × 4.8

P = 4.05

S–16–11. Given triangle DEF, determine the size of angle D in Figure 16–3.

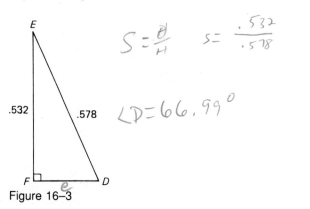
Figure 16–3

S = O/H   S = .532/.578

∠D = 66.99°

S–16–12. Determine the length of e in triangle DEF in Figure 16–3.

∠E A/H = cos ∠E = .532/.578

∠E = 23.01

e = √((.578)² − (.532)²)

E = .226

S–16–13. An airplane flies 1250 miles northeast. How far east is the plane from its original position?

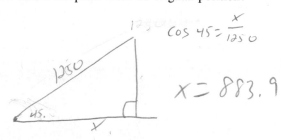

cos 45 = x/1250

x = 883.9

S–16–14. Determine the taper angle of a crankshaft having a taper of 0.75 inch per foot.

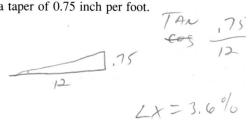

TAN .75/12

∠x = 3.6%

Chapter 16 / Solution of Right Triangles

S-16-15. Six bolt holes are to be drilled on the circumference of a 6-inch hub of a wheel. Determine the straight-line distance between the holes.

INTRODUCTION

It is essential to know trigonometric functions or ratios to solve triangles. Trig functions are found in trig tables or by using a scientific calculator. Some trig functions are as easily found; for example, angles of 30°, 45°, and 60° were determined in Chapter 15. Trig tables, as they are called, have been developed to give the trig function of various angles. These tables give the value of the basic trig functions: sin, cos, and tan. The place value of the function and angles depends on the type of table. For example, some tables are set up with 10′ intervals, some in 1′ intervals, and some tables to the nearest tenth or hundredth of a degree.

Trig tables with angles in decimal form are read the same as degree-minute tables. Minutes are converted to a decimal part of a degree. For example, 38°42′ is read 38 degrees and 42 minutes or, converted to a decimal, $38\frac{42}{60}° = 38.7°$. To convert 38.7° to degrees and minutes do the following:

$$38.7° = 38° + 0.7° \times \frac{60'}{1°}$$
$$= 38° + 7° \times 60'$$
$$= 38° + 42'$$
$$= 38°42'$$

The trig table in this book is set up with 10′ intervals. The table is a four-place table, which means that the values are given to the nearest ten thousandth. The procedure to read this table is the same procedure as that used to read any trig table. Each number represents the ratio of one side of a triangle to another side, depending on which trig ratio and the size of the angle.

EXERCISE 16–1

Use the table of trigonometric functions to find the functions of the angles indicated.

16–1. Find the tangent value of 31°.

Without trig tables, triangles could only be solved by precise construction of the sides and angles.

Example 16–1

PROBLEM: Find the sine value of 38° in the table of trigonometric functions.

SOLUTION: Locate 38° in the table of trigonometric functions. Read the column titled "sin." Thus sin 38° = 0.6157.

Example 16–2

PROBLEM: Find the tangent value of 14°30′ in the table of trigonometric functions.

SOLUTION: Locate 14°30′ in the table of trigonometric functions. Read the column titled "tan"; thus tan 14°30′ = 0.2586.

Example 16–3

PROBLEM: Find the cosine value of 76°20′ in the table of trigonometric functions.

SOLUTION: Locate 76°20′ in the table of trigonometric functions. Since the angle is *greater* than 45°, read the columns from the bottom of the page. Thus you will read the column from the bottom left, titled "cos," so cos 76°20′ = 0.2363.

Remember that when an angle is *less* than 45°, read the table from *left to right* and use the titles at the *top* of the table; when an angle is *greater* than 45°, read the table from *right to left* and use the titles at the *bottom* of the table.

16–2. Find the sine value of 21°.

16-3. Find the cosine value of 30°30'.

.8616

16-4. Find the cosine value of 86°30'.

.06105

16-5. Find the tangent value of 63°50'.

2.0353

16-6. Find the cosine value of 37°20'.

.7951

16-7. Find tan 17°40'.

.31805

16-8. Find sin 38°50'.

.6271

16-9. Find tan 78°10'.

4.7729

16-10. Find sin 45°20'.

.7112

FINDING AN ANGLE GIVEN THE TRIGONOMETRIC FUNCTION

Using the skills learned, but in reverse order, you can determine the degrees of the angle of a given trigonometric function. The examples will show how to locate the degrees of angles of trigonometric functions given the trigonometric function.

Example 16-4

PROBLEM: Find the angle whose sine is .3907.

SOLUTION: Locate the sine value of 0.3907 in the table of trigonometric functions. Then read the column titled "angle" and 23°. The following abbreviations may be used:

$$\sin A = 0.3907$$
$$A = 23°$$

Example 16-5

PROBLEM: Find the angle whose tangent is 2.2113.

SOLUTION: Locate the tangent value of 2.2113 in the table of trigonometric functions. Read the column titled "angle." Since the tangent is greater than 45°, read from the bottom up to locate 65°40'. Thus

$$\tan B = 2.2113$$
$$B = 65°40'$$

Recall that when the tangent value is greater than 1, the angle is greater than 45°.

EXERCISE 16-2

Use the table of trigonometric functions to find the angles given the trigonometric functions.

16-11. Find angle A given $\sin A = 0.1219$.

∠A = 7°

16-12. Find angle X given $\tan X = 1.2954$.

∠X = 52.3°

16-13. Determine angle R when $\cos R = 0.4566$.

∠R = 62.8

16-14. How large is angle W when $\tan W = 20.206$?

∠W = 87.2

16-15. Given $\sin J = 0.6225$, find angle J.

16-16. $\cos P = 0.9528$; $P =$

16-17. $\tan K = 2.9319$; $K =$

16-18. $\sin M = 0.0640$; $M =$

16-19. $\cos N = 0.4488$; $N =$

∠N = 63.3°

16-20. $\sin Y = 0.9644$; $Y =$

INTERPOLATION

Not all function values and angles can be located in the table of trigonometric functions. If a function value or angle cannot be located in the table, the value or angle can be determined through a process called *interpolation*. This process involves locating a value between the given values in the table of trigonometric functions and calculating the desired value. The following examples show how to interpolate to determine an unknown value.

Example 16-6

PROBLEM: Determine the tangent value of 36°24′.

SOLUTION: Locate the angle greater than 36°30′ and the angle less than 36°20′ in the table of trigonometric functions. Then locate the tangents of 36°30′ and 36°20′ in the table of trigonometric functions and arrange the angles and tangents as shown below.

$$10'\begin{bmatrix} \tan 36°30' = 0.7400 \\ 4'\begin{bmatrix} \tan 36°24' = ? \\ \tan 36°20' = 0.7355 \end{bmatrix}? \end{bmatrix}0.0045$$

Then subtract the tangent of the smaller angle (0.7355) from the tangent of the greater angle (0.7400) to find the difference between the tangents.

$$\begin{aligned} \tan \text{ of } 36°30' &= 0.7400 \\ -\tan \text{ of } 36°20' &= 0.7355 \\ &= 0.0045 \end{aligned}$$

Set up a proportion and solve for the unknown value of the tangent.

$$\frac{4'}{10} = \frac{?}{0.0045} \qquad \frac{4'(0.0045)}{10} = ? \qquad \tan 4' = 0.0018$$

Then add the tan value of 4′ (0.0018) to the value of the tangent of 36°20′ (0.7355) to the value of 30°24′ (0.7373).

$$\begin{aligned} \tan \text{ of } 36°20' &= 0.7355 \\ +\tan \text{ of } \quad 4' &= 0.0018 \\ \tan \text{ of } 36°24' &= 0.7373 \end{aligned}$$

Example 16–7

PROBLEM: Determine the cosine of 18°38′.

SOLUTION: Locate the angle greater than 18°30′ and the angle less than 18°40′, the given angle 18°38′, in the table of trigonometric functions. Then locate the cosines of 18°40′ and 18°30′ and arrange the angles and cosines as shown below. Remember that the cosine values decrease as the size of the angle increases.

$$10'\begin{bmatrix} 8'\begin{bmatrix} \cos 18°30' = 0.9483 \\ \cos 18°38' = ? \end{bmatrix}? \\ \cos 18°40' = 0.9474 \end{bmatrix}0.0009$$

Then subtract the cosine of the larger angle (0.9474) from the cosine of the smaller angle (0.9483) to find the difference in value between cosines.

$$\begin{aligned} \cos \text{ of } 18°30' &= 0.9483 \\ -\cos \text{ of } 18°40' &= 0.9474 \\ &= 0.0009 \end{aligned}$$

Set up a proportion and solve for the unknown.

$$\frac{8'}{10'} = \frac{?}{0.0009} \qquad \frac{8'(0.0009)}{10} = ? \qquad \cos 8' = 0.00072$$

Subtract the value of 8′ (0.00072) from the cosine of 18°30′ (0.9483) to find the value of the cosine of 18°38′.

$$\begin{aligned} \cos 18°30' &= 0.9483 - 0.00072 \\ \cos 18°38' &= 0.94758 \\ \cos 18°38' &= 0.9476 \end{aligned}$$

Example 16–8

PROBLEM: Determine the angle whose sine is 0.6894.

SOLUTION: Locate the angle greater than and less than the function given in the table of trigonometric functions. Then locate the angles of the sine of 0.6884 and 0.6905 in the table, and arrange the angles and sines as shown. Remember that sine values increase as the size of the angle increases.

$$10'\begin{bmatrix} ?\begin{bmatrix} \sin 43°30' = 0.6884 \\ \sin \ ? \quad\ = 0.6894 \end{bmatrix}0.0010 \\ \sin 43°40' = 0.6905 \end{bmatrix}0.0021$$

Then subtract 0.6884 from 0.6894 to determine the difference.

$$\begin{aligned} \sin ? \quad &= 0.6894 \\ -\sin 43°30' &= 0.6884 \\ &= 0.0010 \end{aligned}$$

Set up a proportion and solve for the unknown.

$$\frac{?}{10} = \frac{0.0010}{0.0021} \qquad ? = \frac{10'(0.0010)}{0.0021} = 4.76' \text{ or } 5'$$

Add 5′ to 43°30′ to determine the angle whose sin is 0.6894.

$$\begin{aligned} \text{angle whose sin is } 0.6884 &= 43°30' \\ +\text{ angle whose sin is } 0.0010 &= \quad\ 5' \\ \text{angle whose sin is } 0.6894 &= 43°35' \end{aligned}$$

Example 16–9

PROBLEM: Determine the size of angle S whose cosine is 0.1270.

SOLUTION: Locate the cosine greater than and less than that given in the table of trigonometric functions. Arrange the angles and cosines as shown and find the difference. Remember that cosine values decrease as the size of the angle increases.

$$10'\left[?\begin{bmatrix}\cos 82°40' = 0.1276\\ \cos S = 0.1270\\ \cos 82°50' = 0.1248\end{bmatrix}0.0006\right]0.0028$$

Set up a proportion and solve.

$$\frac{?}{10} = \frac{0.0006}{0.0028} \quad ? = \frac{10'(0.0006)}{0.0028} = 2.14' \text{ or } 2'$$

Then, add 2' to 82°40' to determine the degrees in angle S.

$$\cos S = 0.1270$$
$$S = 82°42'$$

EXERCISE 16–3

Use the table of trigonometric functions to find given angles and functions.

16–21. Find the sine of 2°15'.

.0393

16–22. Find the tangent of 19°33'.

TAN = .3551

16–23. Find the cosine of 54°12'.

cos 54°12' = .5850

16–24. Find the tangent of 38°5'.

TAN = .805

16–25. Find the sine of 87°41'.

16–26. Find the cosine of 45°8'.

cos = .6905

16–27. Find the tangent of 87°44'.

TAN = 25.264

16–28. Find the tangent of 8°8'.

16–29. Find A if $\sin A = 0.9850$.

∠A = 80°03'

16–30. Find D if $\cos D = 0.4530$.

∠D = 63.1°

16–31. Find X given $\tan X = 2.2800$.

$\angle X = 66.32°$

16–32. Find P, given $\cos P = 0.8580$.

$\angle P = 30.9°$

16–33. Find Q, if $\sin Q = 0.8809$.

$\angle Q = 61.75°$

16–34. Given that $\cos R = 0.7777$, find R.

$\angle R = 38.95$

16–35. Given that $\tan V = 0.0330$, $V = ?$.

$\angle V = 1.89$

USING A SCIENTIFIC CALCULATOR

The use of a scientific calculator is easier and more efficient than any table of trigonometric functions. Scientific calculators give instantaneous and more accurate trig functions than tables. Scientific calculators have keys for sin, cos, and tan. The other trig functions—cot, sec, and csc—can be obtained by using the reciprocal key.

The decimal degree notation is the standard notation on most electronic calculators. Some scientific calculators have a key that automatically converts angles given in degrees, minutes, and seconds into decimal degrees; the inverse key will reverse this process. Consult your calculator manual to be certain of the proper procedure.

Other calculators have an automatic conversion key from degrees to radians and radians to degrees. Be sure to determine the correct unit before obtaining the trig function.

In some scientific and technical fields, angles are measured in a unit called radians. By definition:

$$1 \text{ radian} = \frac{360°}{2\pi} \text{ degrees or } 57.296° \text{ or } 57.3°$$

One radian is the angle at the center of a circle that corresponds to an arc exactly 1 radian in length. Figure 16–4 shows the radian degree comparison.

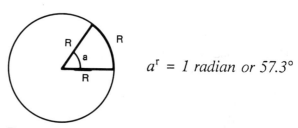

Figure 16–4

$a^r = 1$ radian or $57.3°$

The following ratios demonstrate the radian degree conversion.

$$\frac{\text{angle } a \text{ in radians}}{\text{central angle of the circle}} = \frac{\text{arc length}}{\text{circumference of circle}}$$

$$\frac{a^r}{360°} = \frac{R}{2\pi R}$$

$$\left(\frac{360°}{1}\right)\frac{a^r}{360°} = \frac{R}{2\pi R}\left(\frac{360°}{1}\right)$$

$$a^r = \frac{360°}{2\pi}$$

$$= 57.296°$$

$$= 57.3°$$

Note: A small r above and to the right of the angle indicates radian measure; this symbol may be used similar to the degree symbol°.

$$1^r = 57.3°$$

The following examples will show how to use a scientific calculator to find trig functions and how to find angles when the value of the functions is given. Many calculators operate in both degrees and radians. Be certain to determine which units you are using before using your calculator. Interpolation is done simultaneously with a calculator, so a calculator is easier, faster, and more accurate than tables.

Example 16–10

PROBLEM: Use a scientific calculator to find sin 37.4°.

SOLUTION: Enter 37.4 and press $\boxed{\sin}$ to obtain 0.6073758; round to 0.6074. Thus sin 37.4° = 0.6074.

Example 16–11

PROBLEM: Use a scientific calculator to find tan 76°42′.

SOLUTION: Convert 42′ into a decimal part of a degree by dividing 60 into 42, since 1° = 60′. Thus

$$76°42' = 76\frac{42}{60}° = 76.7°$$

Then enter 76.7′ and press $\boxed{\tan}$ to obtain 4.2302977; round to 4.2303. Therefore, tan 76°42′ = tan 76.7 = 4.2303.

Example 16–12

PROBLEM: Find the angle whose cosine is 0.7880, to the nearest tenth of a degree.

SOLUTION: Enter 0.7880, press $\boxed{\text{inv}}$ and then press $\boxed{\cos}$ to obtain 38. Thus

$$\cos A = 0.7880$$
$$A = 38°$$

Example 16–13

PROBLEM: Find the angle K whose tangent is 0.2600, to the nearest degree, minute, and second.

SOLUTION: Enter 0.2600, press $\boxed{\text{inv}}$, then press $\boxed{\tan}$ to obtain 14.574216. Convert 14.574216 to minutes and seconds as follows:

$$14.574216° = 14° + \left(0.574216° \times \frac{60'}{1°}\right) \quad (1° = 60')$$
$$= 14°34.452966'$$

Then
$$= 14°34'$$
$$+ \left(0.452966' \times \frac{60''}{1'}\right) \quad (1' = 60'')$$
$$= 14°34'27''$$

Thus tan K = 0.2600

$$K = 14°34'27''$$

You should complete problems 16–1 to 16–35 using a scientific calculator to practice finding angles and functions. For the remainder of the book a scientific calculator will be used for calculation to solve all problems.

SOLVING A RIGHT TRIANGLE

Using the skills you have developed using trigonometric functions and angles, you will be able to solve right triangles. To solve a triangle means to find the unknown sides and angles when some of the sides and angles are given. The examples show ways to solve for unknown angles and sides.

Example 16–14

PROBLEM: Given right triangle ABC shown in Figure 16–5, find side a.

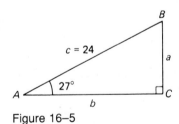

Figure 16–5

SOLUTION: Set up a trig ratio using the given values. This ratio would involve the opposite side a and the hypotenuse c and the given angle A. Use the trig function:

$$\sin A = \frac{\text{opp}}{\text{hyp}}$$

Thus

$$\sin 27° = \frac{a}{24}$$

Find the sin 27° using the table or a calculator. Then solve for a. Remember that the "loop" terms equal side a.

$$0.4540 = \frac{a}{24}$$
$$24(0.4540) = a$$
$$10.896 = a$$

Example 16–15

PROBLEM: Given right triangle ABC shown in Figure 16–6, determine the length of side b.

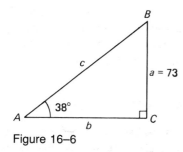

Figure 16–6

SOLUTION: Set up a ratio using the given values. This ratio would involve the opposite side, a, and the adjacent side, b, of angle A. Remember the ratio

$$\tan A = \frac{\text{opp}}{\text{adj}}$$

Thus

$$\tan 38° = \frac{73}{b}$$

Find tan 38° using a calculator. Then solve by multiplying both sides by b.

$$0.7813 = \frac{73}{b}$$
$$b(0.7813) = 73$$

Divide both sides by 0.7813.

$$\frac{b(0.7813)}{0.7813} = \frac{73}{0.7813}$$
$$b = 93.434$$
$$= 93.4$$

Round to the nearest tenth. Remember, accuracy cannot be created through calculation.

Example 16–16

PROBLEM: Given right triangle ABC shown in Figure 16–7, determine the length of side c.

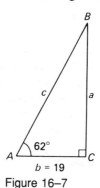

Figure 16–7

SOLUTION: Set up a trig ratio using the given values. This ratio would include the adjacent side, b, and the hypotenuse, c, of angle A. Recall the trig ratio:

$$\cos A = \frac{\text{adj}}{\text{hyp}}$$

Thus

$$\cos 62° = \frac{19}{c}$$

Find cos 62° by using the table or a calculator to complete the ratio. Solve by multiplying both sides of the equation by c.

$$0.4695 = \frac{19}{c}$$
$$c(0.4695) = 19$$

Then divide both sides by 0.4695.

$$\frac{c(0.4695)}{0.4695} = \frac{19}{0.1695}$$
$$c = 40.2556$$
$$= 40.3$$

Round to the nearest tenth.

Example 16–17

PROBLEM: Given the triangle JKL shown in Figure 16–8, find angle K.

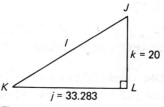

Figure 16–8

Chapter 16 / Solution of Right Triangles

SOLUTION: Set up a trig ratio using the given values. This ratio would include the opposite side, $k = 20$, and the adjacent side, $j = 33.283$, of angle D. Recall that

$$\tan K = \frac{\text{opp}}{\text{adj}}$$

Thus

$$\tan K = \frac{20}{33.283}$$

Then complete the division indicated.

$$\tan K = \frac{20}{33.283}$$
$$= 0.6009$$

Use a calculator to find the angle K, whose tangent is 0.6009. Thus angle $K = 31°$.

Example 16–18

PROBLEM: Solve the right triangle ABC shown in Figure 16–9.

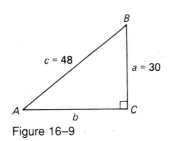

Figure 16–9

SOLUTION: The phrase "solve the right triangle" means to find all of the unknown parts of the right triangle. Therefore, angles A and B and side b must be found. First, find angle A. Set up a trig ratio using the given values: the opposite sides, $a = 30$, and the hypotenuse, $c = 48$. Remember the trig ratio:

$$\sin A = \frac{\text{opp}}{\text{hyp}}$$

Thus

$$\sin A = \frac{30}{48}$$
$$= 0.625$$

Use the table or a calculator to find angle A, whose sine is 0.6250. Thus $A = 38.682° = 38.7°$, or $38°40'$. The sum of angles A and B equals $90°$, so

$$\angle A + \angle B = 90°$$
$$38.7° + \angle B = 90°$$
$$\angle B = 90° - 38.7°$$
$$= 53.3°$$

Then find side b. Set up a trip ratio using the given values: angle A, $38.7°$, and the hypotenuse, 48. Recall the trig function:

$$\cos A = \frac{\text{adj}}{\text{hyp}}$$

Thus

$$\cos 38.7° = \frac{a}{48}$$

Use the table or a calculator to find $\cos 38.7° = 0.7804$. Substitute this into the trig ratio.

$$\cos 38.7° = \frac{b}{48} \quad \text{or} \quad 0.7808 = \frac{b}{48}$$

Solve by multiplying both sides by 84. Thus $b = 37.5$.

$$48(0.7808) = \frac{b}{48}(48)$$
$$37.4784 = b$$
$$b = 37.5$$

Therefore, we have solved triangle ABC by finding all the unknown parts: $\angle A = 38.7°$, $\angle B = 53.3°$, and $b = 37.5$.

Example 16–19

PROBLEM: Given the following parts of a right triangle, PQR—$P = 23°$, $R = 90°$, and side $r = 55$—solve the right triangle.

SOLUTION: First, sketch triangle PQR as shown in Figure 16–10. Find angle Q. The sum of angles P and Q equals $90°$, so

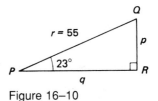

Figure 16–10

$$\angle P + \angle Q = 90°$$
$$23° + \angle Q = 90°$$
$$\angle Q = 90° - 23°$$
$$= 67°$$

Then find side p. Set up a trig ratio using the given values: angle $p = 23°$, and the hypotenuse $r = 55$.

$$\sin A = \frac{\text{opp}}{\text{hyp}}$$
$$\sin 23° = \frac{p}{55}$$

Use the table or a calculator to find sin 23° = 0.3907. Substitute the value into the ratio.

$$\sin 23° = \frac{p}{55}$$
$$0.3907 = \frac{p}{55}$$

Solve by multiplying both sides by 55.

$$55(0.3907) = p$$
$$21.4885 = p$$
$$p = 21.5$$

Then find side q. Set up a trig ratio including the given values: angle $P = 23°$, and the hypotenuse $c = 55$.

$$\cos P = \frac{adj}{hyp}$$
$$\cos 23° = \frac{q}{55}$$

Use the table or a calculator to find cos 23° = 0.9205. Substitute this value into the trig ratio and solve.

$$\cos 23° = \frac{q}{55} \quad \text{or} \quad 0.9205 = \frac{q}{55}$$
$$55(0.9205) = q$$
$$50.6275 = q$$
$$q = 50.6$$

Therefore, we have solved triangle PQR by finding all the unknown parts: $\angle Q = 67°, p = 21.5,$ and $q = 50.6$.

EXERCISE 16–4

Find the value of the unknown sides and angles as indicated.

16–36. Given the triangle ABC, Figure 16–11, find c.

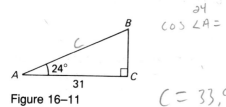

Figure 16–11

$\cos \angle A = \frac{31}{c}$

$c = 33.9$

16–37. Given the triangle DEF, Figure 16–12, find f.

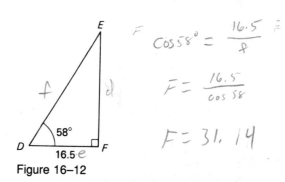

Figure 16–12

$\cos 58° = \frac{16.5}{f}$

$f = \frac{16.5}{\cos 58}$

$f = 31.14$

16–38. Find $\angle G$ in triangle GHI, Figure 16–13.

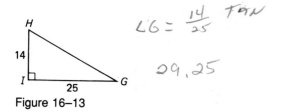

Figure 16–13

$\angle G = \frac{14}{25}$ tan

29.25

16–39. Find the length of side l in triangle JKL, Figure 16–14.

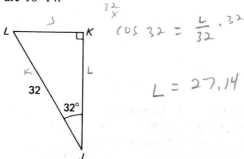

Figure 16–14

$\cos 32 = \frac{L}{32} \cdot 32$

$L = 27.14$

16-40. Find the length of side *n* in triangle *MNO*, Figure 16-15.

16-41. Find ∠*P* in triangle *PQR*, Figure 16-16.

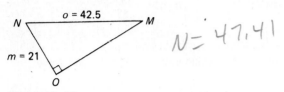

Figure 16-15

N = 47.41

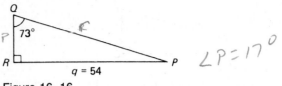

Figure 16-16

∠P = 17°

16-42. Find side *r* in triangle *PQR*, Figure 16-16.

$\sin 73 = \frac{54}{R}$

$R = \frac{54}{\sin 73}$ $R = 56.5$

16-43. Find side *p* in triangle *PQR*, Figure 16-16.

$\tan 73° = \frac{54}{P}$

$P = \frac{54}{\tan 73°}$

$P = 16.5$

16-44. Find the degrees in ∠*S* in triangle *STU*, Figure 16-17.

16-45. Determine ∠*T* in triangle *STU*, Figure 16-17.

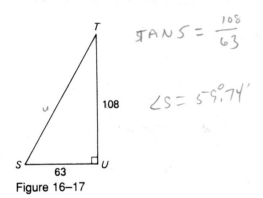

Figure 16-17

$\tan S = \frac{108}{63}$

∠S = 59° 74'

TAN
~~SIN~~ ∠T = $\frac{63}{108}$

~~∠T = 29.~~

∠T = 30.3°

16-46. Find side *u* in triangle *STU*, Figure 16-17.

u = 125.03

16-47. Find ∠*X* in triangle *XYZ*, Figure 16-18.

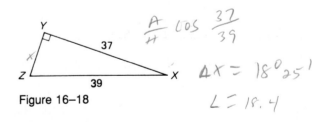

Figure 16-18

$\frac{A}{H}$ $\cos \frac{37}{39}$

∠X = 18° 25'

∠ = 18.4

16-48. Find ∠*Y* in triangle *XYZ*, Figure 16-18.

∠Y = 90°

16-49. Find side *x* in triangle *XYZ*, Figure 16-18.

x = 12.33

16–50. Find the degrees in ∠B, triangle ABC, Figure 16–19.

$\cos = \frac{16}{59}$

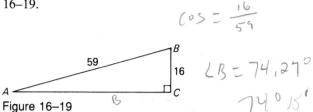

Figure 16–19

∠B = 74.27°

74° 15'

16–51. Find ∠A, triangle ABC, Figure 16–19.

$\sin \frac{16}{59}$ ∠A = 15.73°

16–52. Find side b, triangle ABC, Figure 16–19.

$b = 56.78$

16–53. Given the right triangle, find ∠D, triangle DEF, Figure 16–20.

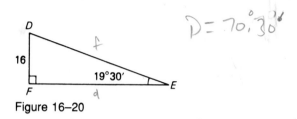

Figure 16–20

D = 70.30°

16–54. Find side d, triangle DEF, Figure 16–20.

$TAN = \frac{16}{D}$

$D = \frac{16}{TAN\ 19.5}$ $D = 45.18$

16–55. Find side f, triangle DEF, Figure 16–20.

$\sin 19.5 = \frac{16}{F}$

$F = \frac{16}{\sin 19.5}$ $F = 47.9$

16–56. Sketch triangle GHI, with ∠I = 90°, H = 21°50', and side g = 175.

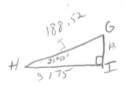

16–57. Find side h, triangle GHI.

$TAN\ 21°50' = \frac{h}{175}$

$h = 70.11$

16–58. Find ∠G of triangle GHI.

68° 10'

16–59. Find side i of triangle GHI.

$i = 188.5$

16–60. Find side j in triangle JKL when side l = 12.48, ∠L = 90°, and ∠J = 83°45'.

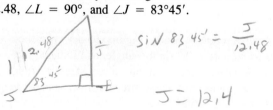

$\sin 83°45' = \frac{j}{12.48}$

$j = 12.4$

APPLICATIONS

The ability to solve right triangles can be most helpful in solving applied problems. In fact, some applied problems can only be solved by using trigonometry. In many problems, the use of trigonometry can help you solve problems more easily and efficiently.

The examples that follow show applications of the solution of right triangle to applied problems. Your imagination will help you find other uses for your ability to solve right triangles.

Example 16–20

PROBLEM: A contractor builds an A-frame cabin 24 feet high with a peak angle of 55°. Find the length of the rafters.

SOLUTION: Draw a sketch of the cabin and indicate the given values as shown in Figure 16–21.

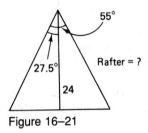

Figure 16–21

Set up a trig ratio using the given values, 27.5°, and the height, 24 feet.

$$\cos A = \frac{adj}{hyp} \quad \text{or} \quad \cos 27.5° = \frac{24 \text{ ft}}{\text{rafter length}}$$

Find the value of cos 27.5° and solve for the rafter length.

$$\cos 27.5° = \frac{24 \text{ ft}}{\text{rafter length}}$$
$$0.8870 = \frac{24 \text{ ft}}{\text{rafter length}}$$
$$\text{Rafter length}(0.8870) = 24 \text{ ft}$$
$$\text{Rafter length} = \frac{24 \text{ ft}}{0.8870}$$
$$= 27.057 \text{ ft}$$
$$= 27.1 \text{ ft}$$

Example 16–21

PROBLEM: The captain of an iron-ore freighter spots a lighthouse at an angle of 2°30′ from his position. The map indicated the light is 158 feet above the surface of the water. How far is the ship from the lighthouse?

SOLUTION: Draw a sketch and label the given quantities as shown in Figure 16–22. Set up a trig ratio

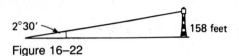

Figure 16–22

using the given values: 2°30′ and the height of the light, 158 feet. Recall that

$$\tan A = \frac{opp}{adj}$$

Thus

$$\tan 2°30' = \frac{opp(158 \text{ ft})}{adj(\text{distance from light})}$$

Find the value of tan 2°30′ or 2.5° $\left(2\frac{30}{60}°\right)$ and solve for the distance.

$$\tan 2°30' = \frac{158 \text{ ft}}{\text{distance}}$$
$$0.0437 = \frac{158 \text{ ft}}{\text{distance}}$$
$$\text{distance} = \frac{158 \text{ ft}}{0.0437}$$
$$= 3615.5606 \text{ ft}$$
$$= 3615 \text{ ft}$$

Example 16–22

PROBLEM: Find the missing dimension x in Figure 16–23.

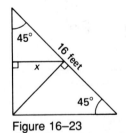 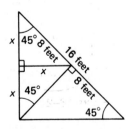

Figure 16–23

SOLUTION: Study Figure 16–23, recalling the properties of a 45° right triangle, and indicate the facts on the sketch. Note that all values of x are equal. Set up a trig ratio using the given values: 45° and the length of one side, 8 feet.

$$\sin A = \frac{\text{opp}}{\text{hyp}}$$

Thus

$$\sin 45° = \frac{\text{opp}(x)}{\text{hyp}} = \frac{x}{8 \text{ ft}}$$

Find the sin 45° and solve for x.

$$\sin 45° = \frac{x}{8 \text{ ft}}$$
$$0.7071 = \frac{x}{8 \text{ ft}}$$

Multiply both sides by 8 feet.

$$x = (8 \text{ ft})(\sin 45°)$$
$$= (8 \text{ ft})(0.7071)$$
$$= 5.6568 \text{ ft}$$

Convert 0.6568 feet to inches.

$$0.6568 \times 12 \text{ in.} = 7.8816 \text{ in.}$$

The inches could be rounded to the nearest eighth inch, which would be $7\frac{7}{8}$ inches. Thus the length of the missing dimension is 5 feet $7\frac{7}{8}$ inches.

Example 16–23

PROBLEM: Find the depth of cut required to mill a square on the end of a $1\frac{1}{2}$-inch shaft, shown in Figure 16–24.

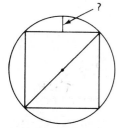
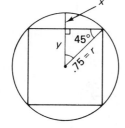

Figure 16–24

SOLUTION: Study Figure 16–24, and enter the given values as shown on the second sketch. Recall the facts of a 45° right triangle. Letter the missing dimension x. Use the letter y as a leg of the right triangle as shown. Thus $x + y = 0.75$ inch. Find the length of y. Set up a trig ratio involving the given values: 45° and the hypotenuse, 0.75 inch.

$$\sin A = \frac{\text{opp}}{\text{hyp}}$$
$$\sin 45° = \frac{y}{0.75}$$

Find the sin of 45° and solve for y.

$$0.7071 = \frac{y}{0.75 \text{ in.}}$$

Multiply both sides of the equation by 0.75 inch.

$$y = (0.75 \text{ in.})(0.7071)$$
$$= 0.530325 \text{ in.}$$
$$= 0.53 \text{ in.}$$

Then solve for x.

$$x + y = 0.75 \text{ in.}$$
$$x + 0.53 \text{ in.} = 0.75 \text{ in.}$$
$$x = 0.75 \text{ in.} - 0.53 \text{ in.}$$
$$x = 0.22 \text{ in.}$$

Example 16–24

PROBLEM: If a missile is launched at 57° at 220 miles per hour, how fast is it traveling horizontally?

SOLUTION: Draw a sketch of the problem as shown in Figure 16–25, and indicate the given values,

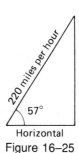

Figure 16–25

220 miles per hour and 57°. Then find the horizontal distance. Recall the trig function that involves the adjacent side and hypotenuse of the given angle.

$$\cos A = \frac{\text{adj}}{\text{hyp}}$$

Substitute the given values and solve for the horizontal speed.

$$\cos 57° = \frac{\text{horizontal}}{220 \text{ miles per hour}}$$
Horizontal = (220 miles per hour)(cos 57°)
Horizontal = (220 miles per hour)(0.5446)
Horizontal = 119.812 miles per hour
Horizontal = 119.8 miles per hour

Chapter 16 / Solution of Right Triangles

EXERCISE 16-5

The following problems will help develop your skills solving applied problems using trigonometry. Draw sketches to better understand the problem you are solving. Refer to the examples if necessary for assistance.

16-61. A triangle is the most rigid of all geometric forms. Determine the length of a diagonal for a cattle gate 10 feet long and 4 feet high.

16-62. A rocket is launched at an angle of 82°30' with a speed of 2800 kilometers per hour. What is its vertical speed in kilometers?

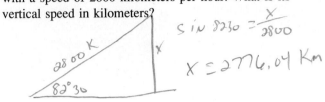

16-63. A roof is to rise 20 feet in a horizontal distance of 32 feet. What is the angle of slope of the roof with the horizon?

16-64. What length of ladder is needed to paint the top of the gable of a house if the gable is 18 feet 6 inches high and the base of the ladder is 10 feet from the house?

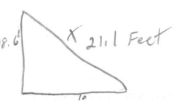

16-65. Find the missing dimension, x, in Figure 16-26.

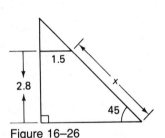

Figure 16-26

16-66. A utility pole is supported by a guy wire 35 feet from the ground. If the wire is inclined at an angle of 49°30' with the ground, how many feet of wire will be needed?

16-67. Point B is 80 air miles north of point A. Point C is 125 air miles east of point B. What is the shortest air mileage between points A and C? (*Note:* Directions are at 90° angles to each other.)

16-68. Find the length, in feet, of the air ventilation shaft of the tunnel shown in Figure 16-27.

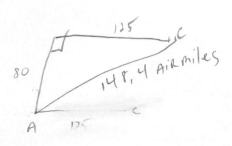

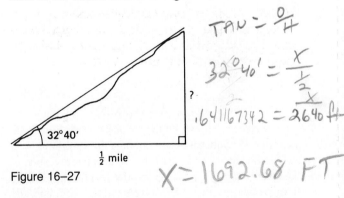

Figure 16-27

16–69. A land surveyor measures 50 feet along a river bank. Then with a transit point directly across the river from the original point, he reads an angle of 76°. How wide is the river?

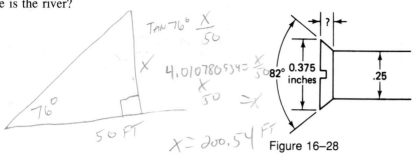

16–70. Find the missing dimension of a flat-head screw shown in Figure 16–28.

Figure 16–28

16–71. Find the taper angle if there is a $\frac{1}{2}$-inch taper in $2\frac{1}{2}$ inches.

16–72. The Washington Monument is 555 feet high. What is the angle of elevation to the top from a distance of 1 mile?

16–73. A freeway shown in Figure 16–29 is constructed through a city at an angle of 68° to the local streets. If the streets are 380 feet apart, what length will the freeway be between intersections of north/south streets?

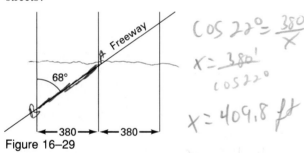

Figure 16–29

16–74. Find the volume of the right prism shown in Figure 16–30.

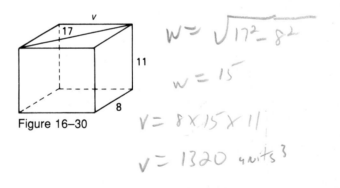

Figure 16–30

16–75. A sheet of metal 25 centimeters wide is bent to form a V-shaped trough. Will the trough have a greater capacity when it is 12 centimeters wide at the top or when it is 10 centimeters deep?

16–76. A helicopter flying at an altitude of 1 mile observes a heliport landing area at an angle of depression of 24°20′. How far, in feet, is the helicopter from the landing area?

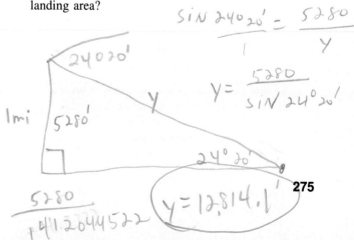

Chapter 16 / Solution of Right Triangles

16–77. A pentagonal flower garden is constructed inside a 16-foot-diameter circle. How much decorative fence must be ordered to fence the garden?

16–78. If the slope of the land is 4°10′, how far will the water be backed up behind a 45-foot dam?

16–79. The end of an 8-centimeter shaft is milled as shown in Figure 16–31. Determine the radius of the smaller circle.

16–80. The 30° arc of a circular curve of a mountain road has a chord of 75 feet. Find the radius of the curve.

Figure 16–31

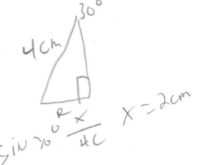

THINK TIME

The procedures used to solve a right triangle can be used to solve everyday problems such as: How high is a building or tree? How far has one sailed, flown, or snowmobiled in a particular direction? Using these procedures to solve problems is limited only to the creative thinking of the person. For example, designers of golf courses determine distances aross natural barriers. Find the shortest distance, from the tee to the hole, in Figure 16–32.

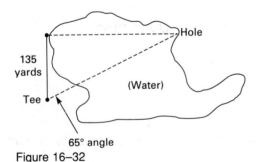

Figure 16–32

PROCEDURES TO REMEMBER

1. To find the trigonometric functions of angles:
   (a) Use a table of trigonometric functions.
   (b) Locate the degree of the given angle.
   (c) Read the appropriate trigonometric function of the angle.
2. To determine the angle of a trigonometric function:
   (a) Use a table of trigonometric functions.
   (b) Locate the given trigonometric value.
   (c) Read the appropriate angle for the given trigonometric function.
3. To determine the trig function of an angle that does not appear in the table of trigonometric functions:
   (a) Locate the angles that are greater and less than the given angle.
   (b) Locate the trig functions that are greater and less than the given angle.
   (c) Subtract the smaller trig function from the greater.
   (d) Set up a trig ratio and solve for the unknown trig value.
   (e) Combine the determined value with the known value.
4. To determine the angle of a trig function that does not appear in the table of trigonometric functions:
   (a) Locate the trig functions that are greater and less than the given function.
   (b) Locate the angles that are greater and lesser than the trig functions.
   (c) Subtract the given trig function from the greater trig function.
   (d) Set up a ratio to find the minutes or parts of a minute represented by the trig function.
   (e) Combine the determined angle with the known angle.

5. To find trig functions using a calculator, enter the angle and press the key for the desired function.
6. To find an angle given a function using a calculator: enter the function, then press the inverse or second function key, and press the given function key.
7. To find the side of a right triangle:
   (a) Set up a trig ratio using the given values.
   (b) Solve the ratio for the unknown side.
   (c) Round accordingly.
8. To find an angle of a right triangle:
   (a) Set up a trig ratio using the given sides to find the value of the trig function.
   (b) Solve the ratio for the trig function.
   (c) Use a table or calculator to find the angle.

CHAPTER SUMMARY

The following definitions and procedures are essential to solve right triangles.

1. *Interpolation* is the process of finding the value of a trig function or angle using known trig functions or angles or a table of trigonometric functions. Determination of the interpolation is not necessary; when using a calculator it is done automatically.
2. The determination of the *arc function* is based on the value of a trigonometric function and the angle it represents.
3. When the tangent value is greater than 1, the angle is greater than 45°.
4. Sine values *increase* as the size of the angle increases.
5. Cosine values *decrease* as the size of the angle increases.
6. Tangent values *increase* as the size of the angle increases.
7. To solve a right triangle means to find all the unknown parts of the right triangle.

CHAPTER TEST

T–16–1. Find the tangent of 76°.

T–16–2. Find the cosine of 32°40′.

T–16–3. Sin 81°20′ =

T–16–4. Find the angle whose cosine is 0.4823.

T–16–5. cos A = 0.7679; A =

T–16–6. Find the cosine of 29°33′.

T–16–7. tan A = 3.6796; A =

T–16–8. Find the length of side *b* in triangle *ABC*, Figure 16–33.

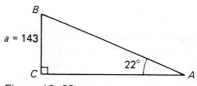

Figure 16–33

T–16–9. Given triangle JKL, Figure 16–34, find angle K.

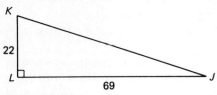

Figure 16–34

T–16–10. Sketch the right triangle and find the length of side a if $A = 37°$, $C = 90°$, and side $c = 303$ centimeters.

T–16–11. Find side s in triangle RST, Figure 16–35.

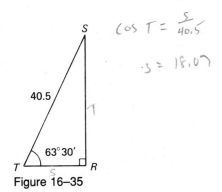

$\cos T = \dfrac{s}{40.5}$

$s = 18.07$

Figure 16–35

T–16–12. Find the length of side t in triangle RST, Figure 16–35.

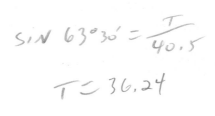

$\sin 63°30' = \dfrac{t}{40.5}$

$t = 36.24$

T–16–13. A 24-foot ladder leans against a wall. Find the angle the ladder makes with the wall if the ladder is $9\dfrac{1}{2}$ feet from the base of the wall.

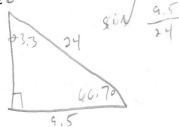

$\sin \dfrac{9.5}{24}$

23.3   24

66.7°

9.5

T–16–14. From point A, a sailboat sails $1\dfrac{1}{2}$ miles due west to point B and then 4 miles south to point C. What is the distance from point A to point C?

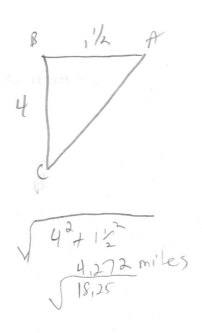

$\sqrt{4^2 + 1\frac{1}{2}^2}$

4.272 miles

$\sqrt{18.25}$

T–16–15. What is the largest square that can be milled from the end of a $1\dfrac{1}{2}$-inch shaft?

17

SOLUTION OF OBLIQUE TRIANGLES

OBJECTIVES

1. To read a trigonometric functions table or use a calculator to determine trig functions of angles greater than 90°.
2. To solve oblique triangles using the law of sines and the law of cosines.
3. To solve applied problems involving oblique triangles.

SELF-TEST

This test will evaluate your ability to solve oblique triangles. If you solve these oblique triangle problems, your trigonometry skills in the solution of oblique triangles should be adequate. Be sure to show your work so that a possible area of concern can be identified.

S–17–1. Find the cos 119°.

S–17–2. Find the sin 169°33′.

S–17–3. Find side a in the oblique triangle ABC in Figure 17–1.

S–17–4. In an oblique triangle, find side d if $\angle D = 118°$, $\angle E$ 27°, and $f = 59.4$.

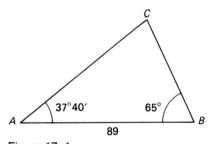

Figure 17–1

S-17-5. Find angle *I* of oblique triangle *GHI* in Figure 17-2.

S-17-6. Find side *k* of triangle *JKL* if side *j* = 426, side *l* = 304, and ∠*K* = 32°40′.

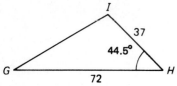

Figure 17-2

S-17-7. Find side *o* in triangle *MNO* in Figure 17-3.

S-17-8. Find ∠*M* in triangle *MNO* in problem S-17-7.

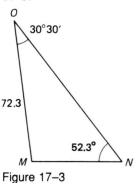

Figure 17-3

S-17-9. From problem S-17-7, find side *m* in triangle *MNO*.

S-17-10. Find the shortest distance across the ravine from point *A* to point *B* in Figure 17-4. Use the sketch as a guide.

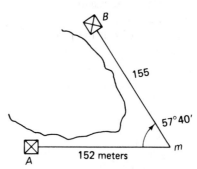

Figure 17-4

INTRODUCTION

Recall that oblique triangles do not contain a right angle. The solution of oblique triangles is more difficult than solving right triangles. The laws of sines and cosines are used to solve oblique triangles. Some oblique triangles can be divided into two or more right triangles for an easy solution. The sum of the interior angles of all triangles is 180° and there are six parts: three sides and three angles.

DEFINITIONS

Two facts concerning sines and cosines of angles greater than 90° are important before the laws of sines and cosines. If an angle is greater than 90° and less than 180°, the sine is equivalent to the sine of 180° minus the given angle. Thus if ∠*A* > 90°, then

$$\sin A = \sin(180° - \angle A)$$

So if $\angle A = 127°$,

$$\sin 127° = \sin(180° - 127°)$$
$$= \sin 53°$$

A calculator will give the correct sine value regardless of the size of the angle.

If an angle is greater than 90° and less than 180°, the cosine is equal to a negative cosine value of 180° minus the given angle. Thus if $\angle A > 90°$, then

$$\cos A = -\cos(180° - \angle A)$$

So if $\angle A = 164°$,

$$\cos 164° = -\cos(180° - 64°)$$
$$\cos 164° = -\cos 16°$$

Again, a calculator will give the correct cosine value regardless of the size of the angle.

Remember, the cosine value of an angle greater than 90° is always negative. Further proof of these concepts may be obtained in more advanced trigonometry texts.

USING THE LAW OF SINES

To solve oblique triangles using the laws of sines and cosines, one of four combinations of values must be given: (1) two angles and one side, (2) two sides and the angle opposite one of the sides, (3) two sides and the angle between the two given sides, and (4) all three sides. Thus at least three of the six parts of the triangle must be given. The specific parts that are given will determine which law will be selected to solve the oblique triangle.

Although it is not essential to know how the law of sines is developed, a simple study of the process will help understand the concept. Study the illustration and the procedure given using triangle ABC in Figure 17–5.

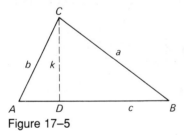

Figure 17–5

Draw a perpendicular from point C to point D on side c to form right triangles, ACD and BCD. State the sine ratios of $\angle A$ and $\angle B$.

$$\sin A = \frac{\text{opposite}}{\text{hypotenuse}} = \frac{k}{b}$$
$$\sin B = \frac{\text{opposite}}{\text{hypotenuse}} = \frac{k}{a}$$

Solve each ratio for k.

$$\sin A = \frac{k}{b} = b \sin A = k$$
$$\sin B = \frac{k}{a} = a \sin B = k$$

Since $k = k$,

$$b \sin A = a \sin B$$

Divide both sides of the equation by $\sin A \sin B$.

$$\frac{b \sin A}{\sin A \sin B} = \frac{a \sin B}{\sin A \sin B}$$
$$\frac{b}{\sin B} = \frac{a}{\sin A}$$

In a similar manner, a perpendicular could be dropped from point A which would result in ratios:

$$\frac{b}{\sin B} = \frac{c}{\sin C}$$

Thus we have developed the law of sines:

$$\frac{a}{\sin A} = \frac{b}{\sin B} = \frac{c}{\sin C}$$

The law of sines states that the ratio of a side of a triangle to the sine of its angle is equal to the ratios of the other sides and their ratios.

The law of sines could also be stated in inverse form:

$$\frac{\sin A}{a} = \frac{\sin B}{b} = \frac{\sin C}{c}$$

Law of Sines

$$\frac{a}{\sin A} = \frac{b}{\sin B} = \frac{c}{\sin C} \quad \text{or}$$

$$\frac{\sin A}{a} = \frac{\sin B}{b} = \frac{\sin C}{c}$$

To solve triangles using the law of sines, two or more ratios are used, and the given quantities are substituted in the ratios. The triangles may be lettered differently and other letters may be used. The following examples show how the law of sines may be used to solve oblique triangles.

Chapter 17 / Solution of Oblique Triangles

Example 17-1

PROBLEM: Find side a in triangle ABC in Figure 17-6.

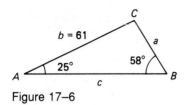

Figure 17-6

SOLUTION: Note the given quantities, $\angle A = 25°$, $\angle B = 58°$, $\angle C = 97°$, and side $b = 61$. Two angles and one side are given; therefore, the law of sines should be used:

$$\frac{a}{\sin A} = \frac{b}{\sin B} = \frac{c}{\sin C}$$

Select two ratios to find side a. The ratios are

$$\frac{a}{\sin A} = \frac{b}{\sin B}$$

Substitute the given values and solve.

$$\frac{a}{\sin 25°} = \frac{61}{\sin 58°}$$

$$\frac{a}{0.4226} = \frac{61}{0.8480}$$

Multiply both sides by 0.4226; thus $a = 30.4$.

$$(0.4226)\frac{a}{0.4226} = \frac{61}{0.8480}(0.4226)$$

$$a = 30.99 \text{ or } 30.4$$

Example 17-2

PROBLEM: Find the size of $\angle C$ given the oblique triangle ABC in Figure 17-7.

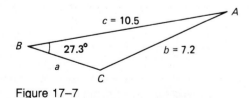

Figure 17-7

SOLUTION: Note the given quantities: $\angle B = 27.3°$, side $b = 7.2$, and side $c = 10.5$. Two angles and one side are given; therefore, the law of sines should be used. Set up two ratios to find the size of $\angle C$. The ratios are

$$\frac{b}{\sin B} = \frac{c}{\sin C}$$

Substitute the given values and solve for angle C.

$$\frac{b}{\sin B} = \frac{c}{\sin C}$$

$$\frac{7.2}{\sin 27.3°} = \frac{10.5}{\sin C}$$

$$\frac{7.2 \sin C}{7.2} = \frac{(10.5)(0.4595)}{7.2}$$

$$\sin C = 0.66966$$

$$\angle C = 42°$$

From the sketch, $\angle C$ is greater than 90°. Using the definition $\sin A = \sin(180° - \angle A)$, substitute as follows:

$$\sin 42° = \sin(180° - 42°)$$
$$= \sin 138°$$

Thus $\angle C = 138°$

Example 17-3

PROBLEM: Solve oblique triangle ABC shown in Figure 17-8.

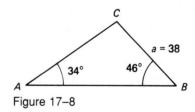

Figure 17-8

SOLUTION: Note that $\angle A = 34°$, $\angle B = 46°$, and side $a = 38$. $\angle C$ and sides b and c are to be found. We are given two sides and one angle; the law of sines should be used to solve the problem. First find $\angle C$.

$$\angle A + \angle B + \angle C = 180°$$
$$24° + 46° + \angle C = 180°$$
$$\angle C = 180 - 34° - 46°$$
$$= 100°$$

Then set up two ratios to find side b. The ratios are

$$\frac{a}{\sin A} = \frac{b}{\sin B}$$

Substitute the given values and solve.

$$\frac{a}{\sin A} = \frac{b}{\sin B}$$

$$\frac{38}{\sin 34°} = \frac{b}{\sin 46°}$$

$$\frac{38}{0.5592} = \frac{b}{0.7193}$$

$$\frac{(38)(0.7193)}{0.5592} = b$$

$$48.8794 = b$$

$$48.9 = b$$

Next set up two ratios to find side c. The ratios are

$$\frac{a}{\sin A} = \frac{c}{\sin C}$$

Substitute the given values and solve for c.

$$\frac{a}{\sin A} = \frac{c}{\sin C}$$

$$\frac{38}{\sin 34°} = \frac{c}{\sin 100°}$$

$$\frac{38}{0.5592} = \frac{c}{0.9848}$$

$$\frac{38 \times 0.9848}{0.5592} = c$$

$$66.9213 = c$$

$$c = 66.9$$

EXERCISE 17–1

Using the law of sines, find the size of angles and the sides as indicated.

17–1. Find side b in triangle ABC, Figure 17–9.

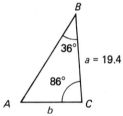

Figure 17–9

17–2. Given $\angle A = 24°$, $\angle C = 65°$, and side $b = 24.3$, determine the length of side c.

17–3. Find the size of $\angle B$ given triangle ABC in Figure 17–10.

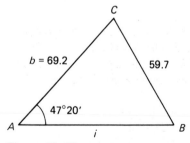

Figure 17–10

17–4. Given the oblique triangle JKL, Figure 17–11, find side k.

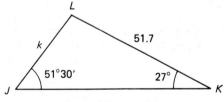

Figure 17–11

Chapter 17 / Solution of Oblique Triangles

17-5. Given $\angle A = 98°6'$, $\angle B = 43°42'$, and side $b = 78.23$, find side a.

17-6. Find side c given the triangle ABC in Figure 17-12.

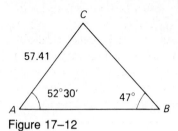

Figure 17-12

17-7. Given the triangle STU, find the size of angle U in Figure 17-13.

17-8. Find the size of $\angle T$ given triangle STU in Figure 17-13.

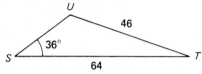

Figure 17-13

17-9. Find side t given triangle STU in Figure 17-13.

17-10. Given the triangle XYZ, find the size of $\angle Z$ in Figure 17-14.

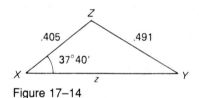

Figure 17-14

17-11. Find the size of $\angle Y$ given triangle XYZ in Figure 17-14.

17-12. Find z given triangle XYZ in Figure 17-14.

17-13. Solve the oblique triangle ABC given $\angle A = 63°$, $\angle B = 42°$, and side $c = 670$. (Remember to draw a sketch.)

17-14. From problem 17-13, find side a.

17-15. From problem 17-13, find side b.

USING THE LAW OF COSINES

The law of sines will not solve all oblique triangles. Triangles with two given sides and the included angle or three given sides can be solved only by using the law of cosines. The law of cosines is stated as follows: The square of any side of a triangle is equal to the sum of the squares of the other two sides minus two times the product of the sides and the cosine of the included angle.

The law of cosines formulas are as follows:

Law of Cosines

$$a^2 = b^2 + c^2 - 2bc \cos A \quad \text{or}$$

$$\cos A = \frac{b^2 + c^2 - a^2}{2bc}$$

$$b^2 = a^2 + c^2 - 2ac \cos B \quad \text{or}$$

$$\cos B = \frac{a^2 + c^2 - b^2}{2ac}$$

$$c^2 = a^2 + b^2 - 2ab \cos C \quad \text{or}$$

$$\cos C = \frac{a^2 + b^2 - c^2}{2ab}$$

The law of cosines is much more difficult to prove than the law of sines. If you desire a proof of the law of cosines, most advanced trigonometry books will contain the proof.

The law of cosines is more difficult to work with than the law of sines; use it only when necessary. The following examples illustrate the law of cosines to solve oblique triangles.

Example 17–4

PROBLEM: Given oblique triangle *ABC* shown in Figure 17–15, find the length of side *c*.

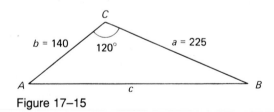

Figure 17–15

SOLUTION: Note the given values: side $a = 225$, side $b = 140$, and $\angle C = 120°$. Two sides and the included angle are given; thus the law of cosines should be used to solve the triangle. Select the law of cosines formula for side *c*.

$$c^2 = a^2 + b^2 - 2ab \cos C$$

or $\sqrt{c^2} = \sqrt{a^2 + b^2 - 2ab \cos C}$

$$c = \sqrt{a^2 + b^2 - 2ab \cos C}$$

To find the cosine value of $\angle C$ (120°), remember that $\angle C$ is greater than 90°. Thus

$$\cos A = \cos(180° - \angle A)$$
$$\cos 120° = -\cos(180° - 120°)$$
$$\cos 120° = -\cos 60°$$
$$\cos 120° = -0.5000$$

Substitute the known values and solve for side *c*.

$$c = \sqrt{a^2 + b^2 - 2ab \cos C}$$
$$= \sqrt{225^2 \ 140^2 - 2(25)(140)(-0.5000)}$$
$$= \sqrt{50,625 + 19,600 + 31,500}$$
$$= \sqrt{101,725}$$
$$= 318.9$$

Example 17–5

PROBLEM: Given the oblique triangle *JKL* shown in Figure 17–16, find the size of $\angle J$.

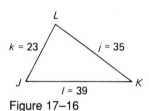

Figure 17–16

SOLUTION: Note the given values: side $j = 34$, side $k = 23$, and side $l = 39$. Three sides are given; thus the law of cosines should be used to solve the triangle. Select the law of cosines formula to solve for $\angle J$. Recall that the square of a side of any triangle is equal to the sum of the squares of the other two sides minus two times the product of those two sides and the cosine of the included angle. Hence the formula for angle *J*:

$$\cos J = \frac{k^2 + l^2 - j^2}{2kl}$$

Substitute known values and solve for cos *J*.

Chapter 17 / Solution of Oblique Triangles

$$\cos J = \frac{(23)^2 + (39)^2 - (35)^2}{2(23)(39)}$$

$$\cos J = \frac{529 + 1521 - 1225}{1794}$$

$$\cos J = \frac{825}{1794}$$

$$\cos J = 0.4599$$

$$\angle J = 62.6°$$

Example 17–6

PROBLEM: Solve oblique triangle ABC, Figure 17–17, $\angle A = 110°$, $b = 24$, and $c = 15$.

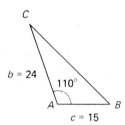

Figure 17–17

SOLUTION: Note the given values: $\angle A$, b and c; solve for $\angle B$, $\angle C$, and a. Two sides and the included angle are given; thus the law of cosines should be used to solve the triangle.

Select the law of cosines formula to solve for a.

$$a^2 = b^2 + c^2 - 2bc \cos A$$

or

$$\sqrt{a^2} = \sqrt{b^2 + c^2 - 2bc \cos A}$$

$$a = \sqrt{b^2 + c^2 - 2bc \cos A}$$

Find the cosine value of $\angle A$ (110°). Recall that $\angle A$ is greater than 90° and use the formula $\cos A = -\cos (180° - \angle A)$:

$$\cos 110° = -\cos(180° - 110°)$$
$$-\cos 70° = -0.3420$$

Then substitute the given values and solve for side a.

$$\sqrt{a^2} = \sqrt{b^2 + c^2 - 2bc \cos A}$$
$$= \sqrt{(24)^2 + (15)^2 - 2(24)(15)(-0.3420)}$$
$$= \sqrt{576 + 225 - 246.24}$$
$$= \sqrt{1047.24}$$
$$= 32.36$$

Note that the multiplication of two negatives $-2(24)(15)$ and -0.3420 resulted in a positive 246.24. Select the law of sines to find $\angle B$.

$$\frac{a}{\sin A} = \frac{b}{\sin B}$$

Find the sine of $\angle A$ (110°) use the formula

$$\sin A = \sin(180° - A)$$
$$\sin 110° = \sin(180° - 110°)$$
$$\sin 70° = 0.9397$$

Substitute the values and solve for sin B.

$$\frac{a}{\sin A} = \frac{b}{\sin B}$$

$$\frac{32.36}{0.9397} = \frac{24}{\sin B}$$

$$(\sin B)(32.36) = (24)(0.9397)$$

$$\frac{\sin B(32.36)}{32.36} = \frac{(24)(0.9397)}{32.36}$$

$$\sin B = 0.6969$$

Thus $\angle B = 44.2°$

Then find $\angle C$.

$$\angle A + \angle B + \angle C = 180°$$
$$\angle C = 180 - 110° - 44.2°$$
$$= 25.8°$$

EXERCISE 17-2

Use the law of sines and the law of cosines to find the angles and sides as indicated.

17-16. Find side b given triangle ABC in Figure 17-18.

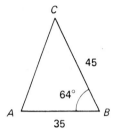
Figure 17-18

17-17. Find the size of angle F in oblique triangle DEF in Figure 17-19.

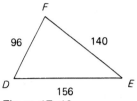
Figure 17-19

17-18. Given oblique triangle GHI, find side i in Figure 17-20.

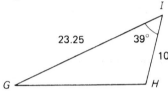
Figure 17-20

17-19. Find side k given oblique triangle JKL in Figure 17-21.

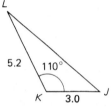
Figure 17-21

17-20. Find the size of angle R in oblique triangle PQR when side $p = 61$, side $q = 65$, and side $r = 76$.

17-21. Find side m in oblique triangle MNO when side $n = 48$, side $o = 28.6$, and $\angle M = 128°$.

17-22. Given triangle TUV, find angle U in Figure 17-22.

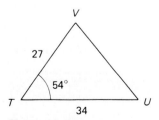
Figure 17-22

17-23. From problem 17-22, find angle V.

Chapter 17 / Solution of Oblique Triangles

17-24. From problem 17-22, find side *t*.

17-25. Given the oblique triangle *GHI* in Figure 17-23, find angle *H*.

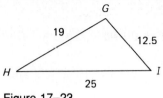

Figure 17-23

17-26. From problem 17-25, find ∠*I*.

17-27. From problem 17-25, find ∠*G*.

17-28. Solve the oblique triangle *ABC* given sides $a = 4.3$, $b = 8.4$, and $c = 5.2$, for ∠*A*. (Remember to draw a sketch.)

17-29. From problem 17-28, find ∠*B*.

17-30. From problem 17-28, find ∠*C*.

APPLICATIONS

The laws of sines and cosines have many uses in industry. Problems concerning navigation, alternating current, and vectors are just a few of the situations where the skills of solving oblique triangles are necessary. The examples that follow illustrate some of these applications.

Example 17-7

PROBLEM: A light plane flew for 85 kilometers and then discovered a course error of 8°30′. After the course correction the destination was reached in 120 kilometers. What should have been the distance if the course error had not been made?

SOLUTION: Make a sketch, Figure 17-24, and

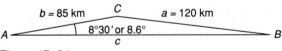

Figure 17-24

identify the given values. Select the laws of sines formula to solve for ∠*B* and solve.

$$\frac{a}{\sin A} = \frac{b}{\sin B}$$

$$\frac{120 \text{ km}}{\sin 8.5°} = \frac{85 \text{ km}}{\sin B}$$

$$(\sin B)(120 \text{ km}) = (85 \text{ km})(0.1478)$$

$$\frac{\sin B(120 \text{ km})}{120 \text{ km}} = \frac{(85 \text{ km})(0.1478)}{120 \text{ km}}$$

$$\sin B = 0.1047$$

$$\angle B = 6°$$

Then find $\angle C$.

$$\angle A + \angle B + \angle C = 180°$$
$$8.5° + 6° + \angle C = 180°$$
$$\angle C = 180 - 8.5° - 6°$$
$$= 165.5°$$

Select the law of sines ratio to solve for c.

$$\frac{a}{\sin A} = \frac{c}{\sin C}$$

$$\frac{120 \text{ km}}{\sin 8.5°} = \frac{c}{\sin 165.5°}$$

Recall:

$$\sin 165.5° = \sin(180° - 165.5°)$$
$$= \sin 14.5° = 0.2504$$
$$\sin 165.5° = \sin 14.5° = 0.2504$$

Thus

$$\frac{120 \text{ km}}{0.1478} = \frac{c}{0.2504}$$

$$\frac{(120 \text{ km})(0.2504)}{0.1478} = c$$

$$203.3 \text{ km} = c$$

The correct course distance should have been 203.3 kilometers.

Example 17–8

PROBLEM: Given Figure 17–25 representing two

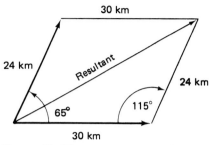

Figure 17–25

concurrent forces acting on a pin connection on a shearing machine, find the resultant.

SOLUTION: Note the given values and the nature of the problem. Two forces acting concurrently form a parallelogram, and the resultant cuts the parallelogram into two oblique triangles. Select the laws of cosines formula to find the resultant.

$$c^2 = a^2 + b^2 - 2ab \cos C$$

$$\text{Resultant} = \sqrt{(\text{force 1})^2 + (\text{force 2})^2}$$
$$- \sqrt{2(\text{force 1})(\text{force 2}) \cos 155°}$$
$$= \sqrt{(24)^2 + (30)^2 - 2(24)(30)(\cos 115°)}$$

Find the cosine of 155° and solve.

$$\cos 115° = -\cos(180° - 115°)$$
$$\cos 115° = -0.4226$$

$$\text{Resultant} = \sqrt{(24)^2 + (30)^2 - 2(24)(30)(-0.4226)}$$
$$= \sqrt{576 + 900 + 608.544}$$
$$= \sqrt{2084.544}$$
$$= 45.6568$$
$$= 45.7 \text{ km}$$

EXERCISE 17–3

Using the laws of sines and cosines, solve the following applied problems. Make a sketch whenever possible.

17–31. From Figure 17–26, showing a triangular-shaped lot, find the length of the missing side.

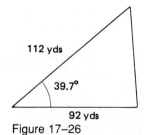

Figure 17–26

17–32. If 18-foot rafters are used for the building shown in Figure 17–27, find the width of the building.

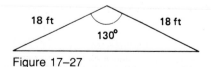

Figure 17–27

Chapter 17 / Solution of Oblique Triangles

17-33. Find the angle of the rafters with the horizon in problem 17-32.

17-34. Find the resultant force in Figure 17-28.

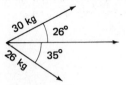
Figure 17-28

17-35. Find the resultant force in Figure 17-29.

17-36. Determine the lengths of the guy wires on the utility pole in Figure 17-30.

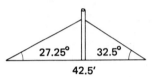
Figure 17-30

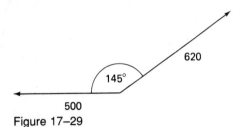
Figure 17-29

17-37. Find the length of the duck-hunting pond shown in Figure 17-31.

17-38. A surveyor measures the length of one side of a tract of land to be 520 rods. From each end of this side, he sights a point and finds the angles to be 74.2° and 65°. Find the length of the other sides.

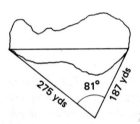
Figure 17-31

17-39. An airplane flies west at 240 miles per hour for 4 hours. A north wind causes the plane to drift to the south at 6.3°. Determine how far the plane must fly to make the correction as indicated in Figure 17-32 and reach the intended destination.

17-40. A helicopter is observed from two ground stations 3560 meters apart. If the angles of elevation are 54° and 28.3°, how high is the helicopter?

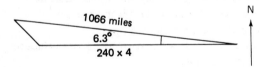
Figure 17-32

17-41. Find the length of an air tunnel to be constructed by the Montana Highway Department for a mountain highway shown in Figure 17-33.

17-42. Two freeways intersect at an angle of 132°. If two cars start from the intersection and travel on the rays of the angle, how far apart will they be in 4 hours if one car is traveling at 55 miles per hour and the other car is traveling at 50 miles per hour?

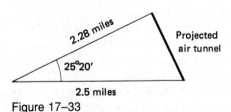
Figure 17-33

17-43. A snowmobile is stuck in a snowbank. A force of 100 pounds is pulling at 25° on the right side of the snowmobile. Another force of 185 pounds is pulling at 25° on the left side. Find the resultant force on the snowmobile.

17-44. A hunter walks in straight lines for 281, 385, and 582 meters. Determine the smallest angle that he makes in his search for game.

17-45. Find the height of the building shown in Figure 17-34.

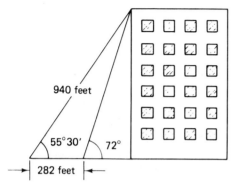

Figure 17-34

THINK TIME

There are many uses for the laws of sines and cosines. Think of ways that you can solve oblique triangles to assist you in your occupation or at home. It may take some imagination to think of uses, but you will find that the laws of sines and cosines will save you time and work.

The example shows how solving oblique triangles can be used to develop recreational and natural resource facilities. A park and recreation director decides to set up a marsh trail shown in Figure 17-35. Because of the low ground, a boardwalk is to be built. She is able to place the walk around the water areas as shown in the sketch. After she has completed the boardwalk around the marsh, she decides to place bridges across the water areas from point D to G and point B to G. Find the length of these bridges. Note that $\angle x = \angle y$.

PROCEDURES TO REMEMBER

1. The procedure for finding the size of an angle of an oblique triangle using the law of sines is:
   (a) Make a sketch of the given triangle.
   (b) Note the given quantities.
   (c) Select two ratios that use the given values to find the side.
   (d) Substitute the given values.
   (e) Solve the ratio for the unknown.
2. The procedure for finding the side of an oblique triangle using the law of cosines is:
   (a) Make a sketch of the triangle to be solved.
   (b) Note the given quantities.
   (c) Select the proper law of cosines formula.
   (d) Find the cosine value of the given angle.
   (e) Substitute the given values.
   (f) Solve for the unknown.
3. The procedure for finding the size of an unknown angle of an oblique triangle, given three sides, using the law of cosines is:

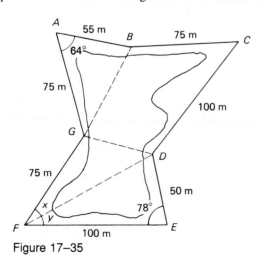

Figure 17-35

Chapter 17 / Solution of Oblique Triangles

(a) Make a sketch of the triangle.
(b) Note the given quantities.
(c) Select the proper law of cosines formula for the unknown angle.
(d) Substitute the given values and solve.

CHAPTER SUMMARY

1. The following definitions are necessary when solving oblique triangles using the laws of sines and cosines.
   (a) Oblique triangles do not contain a right angle.
   (b) If an angle is greater than 90°, the sine of that angle can be determined by the following formula: $\angle A > 90°$; then $\sin A = \sin(180° - \angle A)$.
   (c) If an angle is greater than 90°, the cosine of that angle can be determined by the following formula: $\angle A > 90°$; then $\sin A = -\cos(180° - \angle A)$.
   (d) If the laws of sines and cosines are to be used, one of the following conditions must exist:
       (1) Two angles and one side
       (2) Two sides and the angle opposite one of the sides
       (3) Two sides and the angle between the two sides
       (4) All three sides
   (e) The law of sines states that the ratio of a side to the sine of its angle is equal to the ratios of the other sides and the sines of their angles.
   (f) The law of cosines states that the square of any side of a triangle is equal to the sum of the squares of the other two sides minus two times the product of those sides and the cosine of the included angle.
2. The following formulas can be used to solve oblique triangles:
   (a) Law of sines

   $$\frac{a}{\sin A} = \frac{b}{\sin B} = \frac{c}{\sin C} \quad \text{or}$$
   $$\frac{\sin A}{a} = \frac{\sin B}{b} = \frac{\sin C}{c}$$

   (b) Law of cosines

   $$a^2 = b^2 + c^2 - 2bc \cos A \quad \text{or}$$
   $$\cos A = \frac{b^2 + c^2 - a^2}{2bc}$$
   $$b^2 = a^2 + c^2 - 2ac \cos B \quad \text{or}$$
   $$\cos B = \frac{a^2 + c^2 - b^2}{2ac}$$
   $$c^2 = a^2 + b^2 - 2ab \cos C \quad \text{or}$$
   $$\cos C = \frac{a^2 + b^2 - c^2}{2ab}$$

CHAPTER TEST

T–17–1. Find the sine of 159°.

T–17–2. Find the cosine of 99°27′.

T–17–3. Given the oblique triangle *ABC*, Figure 17–36, find the side *b*.

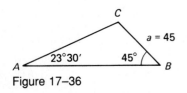

Figure 17–36

T–17–4. Given $\angle A = 65°$, side $a = 14.4$, and side $b = 12.2$, find $\angle B$.

T–17–5. Given the oblique triangle *GHI*, Figure 17–37, find ∠*I*.

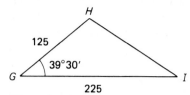

Figure 17–37

T–17–6. Given side *j* = 65, side *k* = 40, and angle *L* = 37°20′, sketch the triangle and find *l*.

T–17–7. Given triangle *MNO* in Figure 17–38, find ∠*O*.

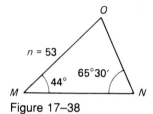

Figure 17–38

T–17–8. From problem T–17–7, find *m*.

T–17–9. From problem T–17–7, find *o*.

T–17–10. Two forces acting at right angles produce a resultant force of 1600 pounds. One of the forces has a magnitude of 960 pounds. Find the other force.

section five: metrication

18

FUNDAMENTALS AND APPLICATIONS OF THE METRIC SYSTEM

OBJECTIVES

1. To identify and understand the fundamental definitions and properties of the metric system.
2. To convert units of measurement (length, area, volume, and weight) to another unit in the metric system.
3. To convert temperature from Celsius to Fahrenheit and from Fahrenheit to Celsius.
4. To convert quantities from the metric system to the English system and from the English system to the metric system.
5. To solve applied problems involving metric measurement.

SELF-TEST

This test will measure your ability to make conversions in the metric system and the English system, plus applied measurement problems. You may use the conversion tables. Show your work so that possible areas of difficulty may be noted.

S–18–1.   521 cm = _____5.21_____ m.

S–18–2.   5 dm = _____.5_____ m.

S–18–3.   5280 m = _____5.28_____ km.

S–18–4.   14 mm = _____1.4_____ cm.

S–18–5.   625 cm² = _____245.83_____ sq in.

S–18–6.   64.255 m² = _____ ha.

S–18–7. 700 miles = 1126.54 km.

S–18–8. 4 lb 3 oz = 1899.4 g.

S–18–9. If top-grade hogs weigh 220 pounds, how many metric tons would 50 top-grade hogs weigh?

= 4.99 m TON

S–18–10. How many kilograms does a 260-pound defensive end weigh?

117.936

S–18–11. 32.3 kg = 32,300 g.

S–18–12. 5 lb = 2.268 kg.

S–18–13. 2.36 metric tons = 5202.86 lb.

S–18–14. 5000 ml = _____ liters.

S–18–15. 5.55 dg = _____ g.

S–18–16. How many grams do 500 milliliters of water weigh?

S–18–17. 45°C = _____ °F.

S–18–18. 210°F = _____ °C.

S–18–19. During a thunderstorm, the temperature dropped from 86°F to 62°F. Find the temperature change on the Celsius scale.

S–18–20. A metal shaper makes a $\frac{3}{64}$-inch cut on a piece of brass. Find the size of the cut in millimeters.

S-18-21. Building codes in a city require the window glass in houses to be at least $\frac{3}{16}$ inch thick. If European windows are used, find the thickness of the glass in millimeters.

S-18-22. A ship loaded with 6600 tons (English) of wheat sails from Duluth, Minnesota to Yokohama, Japan. How many metric tons does the wheat weigh?

S-18-23. A classroom is 30 feet by 40 feet by 12 feet and exchanges air every 3 minutes. How many cubic meters of air is exchanged every hour?

S-18-24. A steel casting weighs 1.5 kilograms. Fifty holes are bored in the casting; 2.75 grams of steel is removed from each hole. Find the weight of the casting after the holes have been bored.

S-18-25. An airplane can fly 1 kilometer on 12 kilograms of fuel. How many metric tons of fuel will be used to fly nonstop from New York to Minneapolis, a distance of 1760 kilometers?

INTRODUCTION

Through the centuries, human beings have reached out to their fellows in an attempt to come to a common understanding concerning life and the world that surrounds us. One such man, L. L. Zamenhof of Poland, observing the diversity of languages, designed an international language he called Esperanto. His attempt to encourage nations to adopt this universal language met with little success.

Gabriel Mouton of France realized the need for a practical decimal system of measurement and developed a base 10 system of measurement in 1670. People liked Mouton's system, but it was not until 1790 until the French Academy endorsed Mouton's system. The French Academy adopted Mouton's system based on the decimal concept and a definite nonchanging value. This definite value was the distance from the North Pole to the equator. The distance was divided into 10 million parts which were called *meters*.

Today, because of the need for an even more precise measure for the meter, the length of the meter is based on the wavelength of the orange light given off by atoms of krypton gas. Other units of measure were derived from the meter so that length, volume, and weight can be uniformly measured.

To ensure precise measurement standards throughout the world, the System International, or SI, international conference established the standards. Nearly all the nations of the world have since adopted the SI metric system. The United States is the only large nation that does not use the metric system. However, much international trade and business is done using the metric system of measure.

DEFINITIONS

There are seven base units of measure in the SI metric system. They are meter (length), kilogram (mass or weight), kelvin (temperature), second (time), ampere (electricity), candela (luminous intensity), and mole (amount of substance). We will study four basic types of metric measure: kilogram, meter, liter, and Celsius degree.

Contrary to English measurement, in the metric system a relationship exists between the basic units of measure and water. Study the relationship shown in Figure 18–1. This relationship between the basic units of measure makes the metric system very logical.

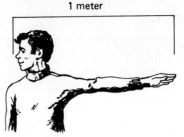

Figure 18–3

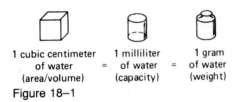

Figure 18–1

As noted before, the metric system is a base 10 or decimal measurement system. This means that each unit may be multiplied or divided by 10 to change a base unit into a more workable unit. Prefixes are added to the base units for identification of the units. When units are *larger* than the base unit, Greek prefixes are used. The most common of these prefixes are *kilo* = 1000, *hecto* = 100, and *deka* = 10. When units are *smaller* than the base unit, Latin prefixes are used. The most common of these prefixes are *deci* = 0.1, *centi* = 0.01, and *milli* = 0.001. Thus a kilometer = 1000 meters and a millimeter = 0.001 meter. Study Figure 18–2 to understand the relationship between metric prefixes and decimal numeration.

Movement of the decimal point to the left or right of a given quantity is all that is needed to change to the next-higher or next-lower unit in the metric system. Naturally, the new unit prefix should be indicated. Thus 89 meters = 890 decimeters. The procedure for making conversions will be shown in the examples.

LENGTH

The base unit of measure for length is the *meter*. A meter may be approximated in the following manner (see Figure 18–3). If an average adult extends his hand out from his shoulder and turns his head away from the extended hand, the distance from the tip of his fingers to the tip of his nose is approximately 1 meter. The thickness of a dime is about the length of a millimeter.

Like other base metric units, we can change the meter into more workable units by either multiplying or dividing by 10. Figure 18–4 shows the various meter units, their names, and the standard abbreviations.

Meter Units		*Comparison to Meter*		*Abbreviation*
1 kilometer	=	1000	meters	km
1 hectometer	=	100	meters	hm
1 dekameter	=	10	meters	dam
1 meter	=	1	meter	m
1 decimeter	=	0.1	meter	dm
1 centimeter	=	0.01	meter	cm
1 millimeter	=	0.001	meter	mm

Figure 18–4 Meter table of equivalents.

It should be noted the dekameter, hectometer, and decimeter are rarely used, whereas the kilometer, meter, and millimeter are commonly used.

Changing from one unit to another may be accomplished simply by moving the decimal point. To better understand the procedure, study the examples.

Example 18–1

PROBLEM: Change 1250 meters (m) into kilometers (km).

SOLUTION: Using the meter chart given, note that 1000 meters = 1 kilometer. Since 1000 meters = 1 kilometer, divide both sides by 1000 meters, thus:

$$\frac{1000 \text{ m}}{1000 \text{ m}} = \frac{1 \text{ km}}{1000 \text{ m}} \quad \text{and} \quad 1 = \frac{1 \text{ km}}{1000 \text{ m}}$$

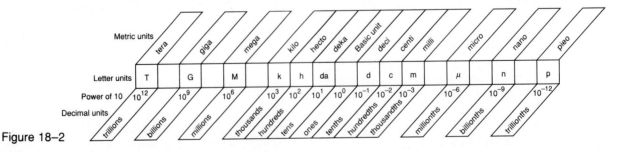

Figure 18–2

Multiply 1250 meters by a fraction equal to 1. (Restudy Example 2–8, if necessary.) In this problem, $\boxed{\dfrac{1\text{ km}}{1000\text{ m}}}$ is the fraction equal to 1. Note that the number $\boxed{1}$, representing the fraction, is used for illustration.

$$\dfrac{\cancel{1250}^{1.25}}{1} \times \boxed{\dfrac{1\text{ km}}{\cancel{1000\text{ m}}_1}} = 1.250\text{ km}$$

An easier procedure to change meters to kilometers is by dividing the given number of meters (1250) by 1000.

$$\dfrac{1250\text{ m}}{1000} = 1.250\text{ km}$$

The easiest procedure is to move the decimal point three places to the left.

km	hm	dam	m	dm	cm	mm
1000	100	10	1	0.1	0.01	0.001
3	2	1				

$$1250\text{ m} = 1\underset{3\ 2\ 1}{\ 2\ 5\ 0}\text{ km}$$

$$1250\text{ m} = 1.25\text{ km}$$

Remember: To change a smaller unit to a larger unit, move the decimal point the appropriate number of spaces to the left.

EXERCISE 18–1

18–1. 15 m = __1500__ cm.

Example 18–2

PROBLEM: Convert 38.8 meters (m) to centimeters (cm).

SOLUTION: Using the meter table, note that 100 centimeters = 1 meter. Multiply 38.8 meters by one: $\boxed{\dfrac{100\text{ cm}}{1\text{ m}}}$.

$$\dfrac{38.8\text{ m}}{1} \times \boxed{\dfrac{100\text{ cm}}{1\text{ m}}} = 3880\text{ cm}$$

An easier procedure to change the meters into centimeters is by multiplying the given number of meters (38.8) by 100.

$$38.8\text{ m} \times 100 = 3880\text{ cm}$$

The easiest procedure is to move the decimal point two places to the right.

km	hm	dam	m	dm	cm	mm
1000	100	10	1	0.1	0.01	0.001
			1	2		

$$38.8\text{ m} = 3\ 8\underset{1\ 2}{\ 8\ 0}\text{ cm}$$

$$38.8\text{ m} = 3880\text{ cm}$$

Remember: To change a larger unit into a smaller unit, move the decimal point the appropriate number of spaces to the right.

18–2. 7 dam = __70__ m.

18–3. 2 mm = __.002__ m.

18–4. 200 cm = __2000__ mm.

Chapter 18 / Fundamentals and Applications of the Metric System

18–5. 53 m = __5.3__ dam.

18–6. 3005 mm = __.003005__ km.

18–7. 2 cm = __.2__ dm.

18–8. 54 hm = __5400__ m.

18–9. 44 cm = __440__ mm.

18–10. 5280 mm = __5.280__ m.

18–11. 12.4 km = __12400__ m.

18–12. 91.5 mm = __.0915__ m.

18–13. 213.6 dm = __21360__ mm.

18–14. 2215 mm = __.02215__ hm.

18–15. 35750 mm = __.035750__ km.

AREA

Recall that area is determined by finding the number of square units contained within an object. Area is expressed in square units such as square meters, square decimeters, and square centimeters. Square units may be abbreviated: for example, a square centimeter is abbreviated as cm^2. The exponent of 2 indicates square units. Common abbreviations for area are as follows: km^2, hm^2, m^2, cm^2, and mm^2.

Land measurement is known by a non-SI metric term: *hectare*. A hectare equals 10,000 square meters, or a square piece of land 100 meters by 100 meters. The following examples illustrate the use of areas in the metric system.

Example 18–3

PROBLEM: Find the number of hectares (ha) in a rectangular plot of land, 1.6 kilometers by 1.2 kilometers.

SOLUTION: Convert kilometers to meters for each of the measurements, since a hectare is equal to 10,000 square meters.

$$1.6 \text{ km} = 1600 \text{ m}$$
$$1.2 \text{ km} = 1200 \text{ m}$$

Using the formula for finding area, $A = lw$, find the area of the plot of land.

$$A = 1600 \text{ m} \times 1200 \text{ m}$$
$$A = 1,920,000 \text{ m}^2$$

Multiply 1,920,000 square meters by $1, \dfrac{1 \text{ ha}}{10,000 \text{ m}^2}$, to find the number of hectares.

$$1,920,000 \times \dfrac{1 \text{ ha}}{10,000 \text{ m}^2} = 192 \text{ ha}$$

Another procedure for converting square meters to hectares is to divide the square meters (1,920,000) by the number of square meters in a hectare (10,000), since 1 hectare = 10,000 square meters.

$$\dfrac{1,920,000 \text{ m}^2}{10,000 \text{ m}^2} = 192 \text{ ha}$$

EXERCISE 18–2

Complete the following area problems.

18–16. 15 dam × 25 dam __37500__ m².
150 m × 250 m =

18–17. 54.5 cm × 35.5 cm __.193__ m².
.545 m × .355 m =

18–18. 0.88 dm × 0.44 dm __38.72__ cm².
8.8 cm × 4.4 cm =

18–19. 35.1 km × 0.5 km __1755__ ha.
351 ha × 5 ha = 1755

18–20. 0.0055 km × 0.0065 km __357,500__ cm².
550 × 650

18–21. A shelf in a closet measures 182 centimeters by 30 centimeters. How many square meters of shelving will be needed for four shelves?
1.82 × .30 m = .546 m²

Chapter 18 / Fundamentals and Applications of the Metric System

[handwritten note at top: why are some x² and other are just x]

18–22. How many hectares are in a wheat farm that is 1.25 kilometers by 0.35 kilometers?

12.5 × 3.5 = 43.75 Hac²

18–23. How many square meters would there be in a table top that is 214 centimeters by 107 centimeters?

2.14 × 1.07 = 2.2898 m²
2.29 m²

18–24. What is the surface area in square centimeters of a cubic box that has a side length of 45 millimeters?

4.5 cm²
A = L × W

18–25. Find the cost of installing carpeting in a room that is 400 centimeters by 350 centimeters if carpet, pad, and installation cost $29.95 per square meter.

4 m × 3.5 m = 14 m² = $419.3

VOLUME

Volume is found by finding the number of cubic units in an object. The basic unit of measure for volume in the metric system is the cubic meter (m³). The exponent 3 indicates cubic measure. A common usage of cubic measurement is found in health occupations, where the term "cc" is used as an abbreviation for cubic centimeters.

The most common unit of measure capacity in the metric system is the liter. Even though the liter is used extensively throughout the world, it is a non-SI unit of measure.

The words "capacity" and "volume" are used interchangeably to mean the same thing. Technically, *volume* is the amount of space within a container, and *capacity* is the amount of substance that a container will hold. The relationship between volume and capacity in the metric system may best be seen in Figure 18–5.

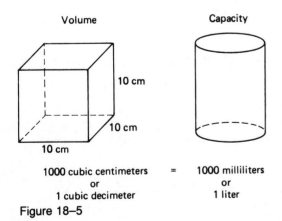

Figure 18–5

Figure 18–6 shows the relationship of a liter in liquid capacity to other units of liquid capacity. Note that the kiloliter, liter, and milliliter are the most commonly used measures.

Liquid Capacity			Volume	
1 kiloliter (kL)	= 1000	liters	= 1000 dm³	= 1.0 m³
1 hectoliter (hL)	= 100	liters	= 100 dm³	= 0.1 m³
1 dekaliter (daL)	= 10	liters	= 10 dm³	= 0.01 m³
1 liter (L)	= 1	liter	= 1 dm³	= 0.0001 m³
1 deciliter (dL)	= 0.1	liter	= 100 cm³	= 0.0001 m³
1 centiliter (cL)	= 0.01	liter	= 10 cm³	= 0.00001 m³
1 milliliter (mL)	= 0.001	liter	= 1 cm³	= 0.000001 m³

Figure 18–6 Liter table of equivalents.

Changing from one unit to another unit with a liter as the base unit can be accomplished by moving the decimal point to the appropriate number of spaces to the right or left. To change a larger unit (liter) to a smaller unit (deciliter), move the decimal point to the right the appropriate number of places (one). To change a smaller unit (liter) to a larger unit (dekaliter), move the decimal point to the left the appropriate number of places (one). Study the following examples that show the procedures.

Example 18–4

PROBLEM: Convert 17 liters (L) to milliliters (mL).

SOLUTION: Using the liter chart, note that 1000 mL = 1 L. Multiply 17 liters by 1 or

$$\frac{17 \text{ L}}{1} \times \frac{1000 \text{ mL}}{1 \text{ L}} = 17{,}000 \text{ mL}$$

Another procedure for changing liters to milliliters is by multiplying the given number of liters (17) by 1000.

17 L × 1000 = 17,000 mL

The simplest solution, however, is to move the decimal point three places to the right.

kL	hL	daL	L	dL	cL	mL
1000	100	10	1	0.1	0.01	0.001

$$17 \text{ L} = 17.000 \text{ mL}$$

$$17 \text{ L} = 1700 \text{ mL}$$

Example 18–5

PROBLEM: Find the volume in cubic centimeters and liters of the aquarium shown in Figure 18–7.

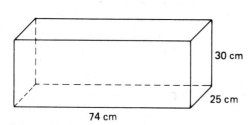

Figure 18–7

SOLUTION: Find the volume of the aquarium. Recall the formula for finding volume: volume (V) = length (l) × width (w) × height (h). Thus

$$\begin{aligned} V &= lwh \\ &= (74 \text{ cm})(25 \text{ cm})(30 \text{ cm}) \\ &= 55{,}500 \text{ cm}^3 \end{aligned}$$

Convert cubic centimeters to milliliters. Recall that 1 cubic centimeter = 1 milliliter.

Thus $55{,}500 \text{ cm}^3 = 55{,}500 \text{ mL}$

Then convert 55,500 milliliters to liters. (*Note:* 1000 milliliters = 1 liter.) Thus

$$55{,}500 \text{ mL} = 55.500 \text{ L}$$

$$55{,}500 \text{ mL} = 55.5 \text{ L}$$

EXERCISE 18–3

Solve the following as indicated, using a conversion table.

18–26. 77 L = __77000__ mL.

18–27. 25 cc = __25__ cm³.   *centimeter cube.*

18–28. 3.8 dm³ = __38__ cm³.

18–29. 55 mL = __5.5__ cL.

18–30. 0.66 cm³ = __.066__ dm³.

18–31. 8880 mL = __8.88__ L.

Chapter 18 / Fundamentals and Applications of the Metric System

18–32. 747,720 mm³ = __747.720__ m³.

18–33. 76.7 cc = __76.7__ cm³.

18–34. 0.0075 dm³ = __.75__ mm³.

18–35. 3836 dL = __.3836__ kL.

18–36. 1.38 m³ = __1980__ mm³. ×3×3 *(not cubed)*
  __1.38 × 10⁹__

18–37. 5001 L = __50.01__ hL.

18–38. 0.0076 cm³ = __.076__ mm³.

18–39. 3.27 mL = __.00327__ L.

18–40. Find the capacity in liters of a gasoline tank 5.5 m long, 2.5 m wide, and 2 m deep.

18–41. Find the volume in cubic centimeters of a soft-water container 2.1 m long, 24 cm wide, and 26 mm deep.

24 × 210 × 2.6 = 13,104 cm³

18–42. Find the volume in cubic meters of a cube with one side 340 cm long.

3.4
= 39.304 cm.

18–43. An excavation for a building is to be 55 m by 40 m by 15 m. What is the cost of excavation at $4.85 per cubic meter?

33000 × 4.85 = $160,050

Section Five / Metrication

18-44. A plastic underlayment sheet for a bike path is 10 km by 2 m by 4 mm. How many cubic meters of plastic are in the sheet?

10,000 × 2 × .004 = 80 cm³

18-45. A swimming pool 25 m long and 10 m wide has an average depth of 1.5 m. How many kl of water will the pool hold?

375 × = 375 KL

18-46. How many 500-milliliter containers of honey could be filled from 100 hectoliters?

500)10,000,000 = 20,000 containers

18-47. How many cubic meters of blacktop are needed to make a driveway 10 m long, 5 m wide, and 12 cm thick?

10 × 5 × .12 = 6 m³

18-48. If rain fell at the rate of 2 mm per minute, how many minutes would it take to reach 0.1 m?

200 m = 1000 50 min

18-49. Find the volume in liters of a rectangular water cooling tank 2 m by 20 dm by 28 cm.

1.12 L

18-50. A patient was given 20 cc of a medicine in 500 ml of water. What percent of the solution was medicine?

$\frac{20}{500} = 4\%$

MASS

Mass is defined as the quantity of material contained in a given body. *Weight* is the measure of the force of gravity on a given body. These are the technical definitions of mass and weight; however, they are frequently used to mean the same thing. For example, on the planet Earth, the terms "mass" and "weight" may be used interchangeably because the Earth's gravity affects objects in a similar manner. However, on the moon, a mass weighs about $\frac{1}{6}$ of its weight on the Earth.

This is due to the moon's gravity, which is about $\frac{1}{6}$ of the Earth's gravity. For our discussion we assume that mass or weight may be used similarly.

The basic SI metric unit for measuring mass is the kilogram. An average-sized paper clip weighs about 1 gram. Thus a kilogram is about the weight of 1000 paper clips.

The relationship between volume capacity and weight in the metric system can be seen in Figure 18-8. The following should be noted:

$$1 \text{ cm}^3 = 1 \text{ mL} = 1 \text{ g}$$
$$1000 \text{ cm}^3 = 1 \text{ L} = 1000 \text{ g}$$
$$1 \text{ dm}^3 = 1 \text{ L} = 1 \text{ kg}$$

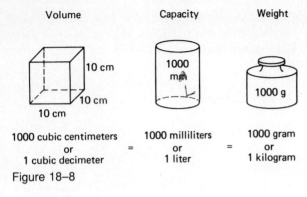

1000 cubic centimeters or 1 cubic decimeter = 1000 milliliters or 1 liter = 1000 gram or 1 kilogram

Figure 18–8

Figure 18–9 shows the relationship of a kilogram of weight to other units of weight, volume, and capacity in the metric system.

For water

Weight		Liquid Capacity		Volume	
1000 kilograms (kg) = 1 metric ton (t)	=	1	kL	= 1	m³
1 kilogram (kg) = 1000 grams	=	1	L	= 1	dm³
1 hectogram (hg) = 100 grams	=	0.1	L	= 100	cm³
1 dekagram (dag) = 10 grams	=	0.01	L	= 10	cm³
1 gram (g) = 1 gram	=	1	mL	= 1	cm³
1 decigram (dg) = 0.1 gram	=	0.1	mL	= 1	mm³
1 centigram (cg) = 0.01 gram	=	0.01	mL	= 0.1	mm³
1 milligram (mg) = 0.001 gram	=	0.001	mL	= 0.01	mm³

Figure 18–9 Kilogram table of equivalents.

It should be noted that 1000 kilograms = 1 metric ton = 1 kiloliter = 1 cubic meter. The procedures for changing units in the metric weight system are the same as those shown previously for changing other metric units. Study the examples that show the procedures.

Example 18–6

PROBLEM: Convert 2700 milligrams (mg) to grams (g).

SOLUTION: Using the kilogram table of equivalents, 1000 milligrams = 1 gram. Multiply 2700 by 1:

$$\frac{2700 \text{ mg}}{1} \times \frac{1 \text{ g}}{1000 \text{ mg}} = 2.7 \text{ g}$$

We could change the milligrams into grams by dividing the milligrams (2700) by 1000.

$$\frac{2700 \text{ mg}}{1000} = 2.7 \text{ g}$$

The easiest way to change mg to g is to move the decimal point three places to the left.

kg	hg	dag	g	dg	cd	mg
1000	100	10	1	0.1	0.01	0.001

2700 mg = 2.700 g

2700 mg = 2.7 g

Example 18–7

PROBLEM: What is the weight in kilograms of gasoline in a rectangular tank 1.3 m long, 8 dm wide, and 42 cm deep? (Gasoline weighs 0.8 times as much as water.)

SOLUTION: Change the measurements to cm: 1 m = 100 cm and 1 dm = 10 cm. Thus

$$1.3 \text{ m} = 130 \text{ cm}$$
$$8 \text{ dm} = 80 \text{ cm}$$
$$42 \text{ cm} = 42 \text{ cm}$$

Find the volume of the tank using the formula $V = lwh$.

$$V = (130 \text{ cm})(80 \text{ cm})(42 \text{ cm})$$
$$V = 436{,}800 \text{ cm}^3$$

Change cubic centimeters to kilograms using the formula 1 kg = 1000 cm³.

$$\frac{436{,}800 \text{ cm}^3}{1} \left(\frac{1 \text{ kg}}{1000 \text{ cm}^3}\right) = 436.8 \text{ kg}$$

Multiply to find the weight of the gasoline, which is 0.8 times the weight of water.

$$(436.8 \text{ kg})(0.8 \text{ weight of gasoline}) = 349.44 \text{ kg}$$

Thus the tank contains 349.44 kilograms.

EXERCISE 18-4

The following problems will give you the opportunity to develop your skills with weight measurements. Use any tables you need to solve the problems.

18-51. Convert 3.8 kilograms to grams.

3,800

18-52. Convert 520 milligrams to grams.

.52

18-53. Convert 81.3 decigrams to milligrams.

8130 g   8130 mg

18-54. Convert 7.8 centigrams to grams.

.078

18-55. Convert 57.3 dekagrams to milligrams.

18-56. 7886 mg = _____ g.

18-57. 85.6 dg = _____ g.

18-58. 45,000 g = _____ kg.

18-59. 800 dag = _____ kg.

18-60. 3804 kg = _____ metric tons.

18-61. 84 g = _____ cg.

18-62. 81 dg = _____ mg.

18–63. 8.08 kg = _____ cg.

18–64. 5.54 tons = _____ kg.

18–65. Find the sum of the following weights: 4.3 g, 4.9 cg, 781 mg, and 5.05 cg.

18–66. Find the sum of the following: 0.83 kg, 0.38 kg, 33.3 g, 5050 mg, and 4.4 dg.

18–67. What is the weight in kg of 34 castings that weigh 386 hg each?

18–68. An irregular-shaped plastic container weighed 1400 grams when empty. When filled with water it weighed 5.4 kilograms. Find the volume of the container in cm^3.

18–69. If there are 8 grams of polluting substance per liter of sewage, how many kilograms of polluting matter remain per 700 kiloliters?

18–70. Find the weight in metric tons of 54 pieces of plastic weighing 85.6 kilograms each?

18–71. A chemist has a 2-liter beaker of an aqueous solution containing 550 grams of sugar. The air evaporates 300 cubic centimeters of the solution. What is the weight of the remaining solution?

18–72. A brick measures 5.8 centimeters by 9.5 centimeters by 20.2 centimeters. Find its weight in kilograms (to the nearest tenth) if bricks weigh 2.4 grams per cubic centimeter.

18–73. Determine the weight of the water in metric tons in a swimming pool 25 m by 10 m by 2 m.

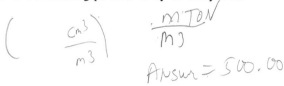

18–74. A beaker holds 500 ml of a salt solution containing 15 g of salt. How much salt must be added to make it a 10% solution?

18–75. Determine the weight of a solid piece of brass 95 mm thick, 5 cm wide, and 3 m long. (Brass weighs 8.2 kg per cubic dm.)

TEMPERATURE

The basic SI unit for measuring temperature is the *kelvin* (K) scale. The kelvin scale is used by scientists and is based on a scale from absolute zero where water freezes at 273.15°K and boils at 373.15°K. Since this scale is not very practical, the Celsius (C) or centigrade scale is used in the metric system. The Celsius scale, based on the kelvin system, indicates water freezing at 0 degrees and boiling at 100 degrees. Thus 1 degree Celsius is equal to 1 degree Kelvin. Figure 18–10 shows the relationships of the three types of measurement scales.

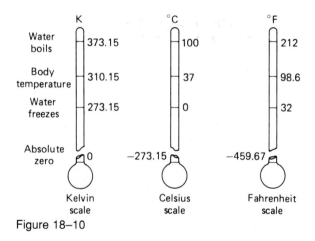

Figure 18–10

The United States may someday adopt the Celsius system of measuring temperature to replace the Fahrenheit temperature measurement. Thus it is necessary to be able to convert from Fahrenheit to Celsius and Celsius to Fahrenheit. The United States is the largest country in the world that does not use the Celsius temperature system. The following temperature conversion formulas will be helpful.

From Celsius to Fahrenheit: $F = 1.8C + 32°$

From Fahrenheit to Celsius: $C = \dfrac{5}{9}(F - 32°)$

The following examples show how the formulas are used.

Example 18–8

PROBLEM: Convert 40°C to °F

SOLUTION: Using the formula $F = 1.8C + 32°$, substitute 40° for C and solve the formula for the unknown, F.

$$F = 1.8C + 32°$$
$$= 1.8(40°) + 32°$$
$$= 72° + 32°$$
$$= 104°$$

Example 18–9

PROBLEM: A chef has an oven with a Celsius thermostat. A recipe calls for a cake to be baked at 320°F. Find the Celsius setting.

SOLUTION: Use the formula $C = \dfrac{5}{9}(F - 32°)$. Substitute 320° for F and solve for the unknown, C.

$$C = \dfrac{5}{9}(320° - 32°)$$
$$= \dfrac{5}{\underset{1}{9}}\left(\dfrac{\overset{32°}{288°}}{1}\right)$$
$$= 160°$$

Thus the Celsius thermostat setting would be 160°.

Chapter 18 / Fundamentals and Applications of the Metric System

EXERCISE 18–5

Converting from one measurement to another is simple when the correct formulas are used. Use the formulas to convert the following temperatures.

18–76. 20°C = _____ °F.

$9/5 (20 + 32)$

18–77. 28°C = __82.4__ °F.

$F = \left(\frac{9}{5} \times 28\right) + 32$

18–78. 98.6°F = __37__ °C.

$C = 5/9 (98.6 - 32)$

18–79. 100°F = __37.8__ °C.

$C = 5/9 (100 - 32)$

18–80. 212°C = __413.6__ °F.

$F = \left(\frac{9}{5} \times 212\right) + 32$

18–81. 62°F = __16.7__ °C.

$C = 5/9 (62 - 32)$

18–82. 45°F = __7.2__ °C.

$C = 5/9 (45 - 32)$

18–83. −30°F = __−34.4__ °C.

$C = 5/9 (\overline{}^{30} F + 32) = -62$

18–84. 72°F = __22.2__ °C.

$C = 5/9 (72 - 32)$
$40 =$

18–85. 85°F = __29.4__ °C.

$C = 5/9 (85 - 32)$

18–86. Sleet occurs when the temperature is about 27°F. What is the Celsius equivalent to this temperature?

18–87. Wrought iron melts at 1550°C. What temperature is this on the Fahrenheit scale?

18–88. The temperature in the oxyacetylene process of welding sometimes reaches 5800°F. What is the temperature on the Celsius scale?

18–89. The temperature dropped from 20°F to −10°F in 5 hours. What would the change of temperature be in degrees Celsius?

18–90. At what temperature is the reading on the Celsius thermometer the same as the reading on the Fahrenheit thermometer?

CONVERTING BETWEEN THE METRIC AND ENGLISH SYSTEMS

As business and industry continues to become international, it is valuable to understand and use the metric system. Thus it is essential to make conversions from one system to the other. Today there is a need to be able to work efficiently with both systems and to convert measurements from one system to the other. Electronic calculators, conversion charts, and other devices aid in the conversion process.

You may find it necessary to convert measurements from one system to the other, so it is important for you to learn the few common equivalents shown in Figure 18–11.

English to Metric		Metric to English	
1 inch	= 2.54 cm	1 cm	= 0.3937 inch
1 yard	= 0.9144 m	1 m	= 1.0936 yards
1 mile	= 1.609 km	1 km	= 0.6213 mile
1 ounce (dry)	= 28.349 g	1 g	= 0.0352 ounce
1 pound (dry)	= 0.4536 kg	1 kg	= 2.2046 pounds
1 ounce (liquid)	= 29.75 mL	1 mL	= 0.0338 ounce
1 quart (liquid)	= 0.946 L	1 L	= 1.0568 quarts

Figure 18–11 Common English and metric equivalents.

The basic conversions given, along with your understanding of the metric and English systems, should allow you to move from one system to the other. You may find it helpful to memorize some of the commonly used conversion factors. Additional conversion factors are stated in Figures 18–12 to 18–15.

English to Metric		Metric to English	
1 in.	= 2.54 cm	1 mm	= 0.03937 in.
	= 25.4 mm	1 cm	= 0.3937 in.
1 ft	= 30.48 cm		= 0.0328 ft
	= 0.3048 m	1 m	= 39.37 in.
1 yd	= 91.44 cm		= 3.2808 ft
	= 0.9144 m	1 km	= 0.62136 mile
1 mile	= 1609.344 m		= 3280.8 ft

Figure 18–12 Length equivalents.

English to Metric		Metric to English	
1 sq in.	= 6.4516 cm²	1 cm²	= 0.155 sq in.
1 sq ft	= 0.09290 m²	1 m²	= 10.7639 sq ft
1 sq yd	= 0.83613 m²	1 hectare	= 2.471 acres
1 acre	= 0.4047 ha		

Figure 18–13 Area equivalents.

English to Metric		Metric to English	
1 cu in.	= 16.387 cm³	1 cm³	= 0.061 cu in.
1 cu ft	= 28.317 dm³	1 dm³	= 61.0237 cu in.
	= 0.0283 m³		= 0.0353 cu ft
1 cu yd	= 0.7646 m³	1 m³	= 35.3147 cu ft
			= 1.3079 cu yd

Figure 18–14 Volume equivalents.

English to Metric		Metric to English	
1 pound	= 0.4536 kg	1 kilogram	= 2.2406 lbs
1 short ton	= 907.2 kg	1 gram	= 0.0352 oz
1 long ton	= 1016 kg	1 metric ton	= 2204.6 lbs

Figure 18–15 Weight and mass equivalents.

The examples will give you an opportunity to study the conversion procedures and to practice making conversions.

Example 18–10

PROBLEM: How long is a 100-yard football field in meters?

SOLUTION: Using the conversion fact 1 yard = 0.9144 meter, multiply by $\dfrac{0.9144 \text{ m}}{1 \text{ yd}}$.

$$\left(\dfrac{100 \text{ yd}}{1}\right)\left(\dfrac{0.9144 \text{ m}}{1 \text{ yd}}\right) = 91.44 \text{ m}$$

Thus a 100-yard football field is 91.44 meters long.

Example 18–11

PROBLEM: In the United States, the speed limit for automobiles is 65 miles per hour. What is the speed limit in kilometers per hour?

SOLUTION: Use the conversion fact 1 mile = 1.609 kilometers. Thus $\dfrac{1.609 \text{ kmph}}{1 \text{ mph}}$.

$$\left(\dfrac{65 \text{ mph}}{1}\right)\left(\dfrac{1.609 \text{ kmph}}{1 \text{ mph}}\right) = 104.585 \text{ kmph}$$

Thus 65 miles per hour is equal to 104.6 kilometers per hour.

Example 18–12

PROBLEM: The capacity of a gasoline tank in a German sportscar is 42 liters. How many gallons is this to the nearest tenth of a gallon?

SOLUTION: Using the conversion facts 1 liter = 1.0568 quarts and 4 quarts = 1 gallon, set up the procedures and solve for gallons.

$$\left(\dfrac{42 \text{ L}}{1}\right)\left(\dfrac{1.0568 \text{ qt}}{1 \text{ L}}\right)\left(\dfrac{1 \text{ gal}}{4 \text{ qt}}\right) = 11.0964 \text{ gal}$$

Thus 42 liters is equal to about 11.1 gallons.

Example 18–13

PROBLEM: A French chef is to prepare 14 servings of filet mignon for a special party. Each filet should weigh about 250 grams. How many pounds of steak will the chef need to order from his suppliers?

SOLUTION: Change grams to kilograms by moving the decimal point three places to the left.

$$250 \text{ g} = .250 \text{ kg}$$

$$250 \text{ g} = 0.25 \text{ kg}$$

Then find the total amount of meat needed in kilograms.

14 servings × 0.25 kg per serving = 3.5 kg of meat

Using the conversion fact 1 kilogram = 2.2046 pounds, set up the procedure and solve for pounds.

$$\left(\dfrac{3.5 \text{ kg}}{1}\right)\left(\dfrac{2.2046 \text{ lb}}{1 \text{ kg}}\right) = 7.7161 \text{ lb of meat}$$

Thus the chef should order 7.72 pounds of meat from his suppliers to serve 14 people 250-gram filets.

EXERCISE 18–6

Conversion from metric to English and English to metric is not difficult when the correct facts are used and the proper mathematical operations are applied. Practice making conversions using the correct conversion formulas.

18–91. 1 ft = __30.48__ cm.

18–92. 1 mile = __1609.344__ m.

18–93. 8.3 km = __5.16__ miles.

8.3 km × .6213 miles = 5.16

18–94. 105 mm = __4.133__ in.

1 cm = .3937 in
1 mm = .03937
.3937

18–95. 55 m = _____ yd.

1 yd = .9144 m
55 × .9144

18–96. 24 in. = __609.6__ mm.

24 × 25.4

18–97. 66 sq yd = __55.2__ m².

66 × .8361

18–98. 48 cm² = __7.44__ sq in.

48 × .155 in² / 1 cm²

18–99. 66.2 mm² = __.103__ sq in.

66.2 mm² × 1 cm²/100 mm² × 1 in²/6.4516 cm²

66.12 / 645.6

18–100. What is the size in centimeters of a 12-inch by 12-inch tile?

12 in = 2.54 = 30.48

18–101. Find the area of a 12-inch by 12-inch tile in square centimeters.

929.03

18–102. $\frac{1}{2}$ gal = __1.892__ L.

2 quarts = 2 × .946

18–103. 25 cu in. = __410__ cm³.

25 × 16.387

18–104. 1 L = __35.2__ oz.

1 = 1000 grams   page 306
1 gram = .0352   ?.3/1

18–105. 0.35 kg = _____ oz.

18–106. One-half kilogram of nails costs $0.98. What is the cost per pound?

18–107. 25,000 cu in. = _____ cm³.

18–108. A 12-ounce container of frozen orange juice is equal to how many ml?

18–109. How many cubic meters of earth will need to be excavated for a house 80 feet by 30 feet and 4 feet deep?

18–110. A contractor needs 150 cubic yards of concrete for a job. How many cubic meters is this?

18–111. 220 lb = __99.8__ kg.

18–112. 53 kg = _____ lb.

18–113. A 100-pound sack of potatoes is equal to how many kilograms?

18–114. Number 1 spring wheat weighs 60 pounds per bushel (1.244 cubic feet). How many kilograms does the wheat weigh per cubic decimeter?

18–115. A liter of milk equals how many ounces?

18–116. A spark gap is 1.5 millimeters. How many thousandths of an inch is this?

18-117. A closet shelf is 1.83 meters long by 61 centimeters wide. How long and wide is the shelf to the nearest inch?

18-118. A garden is 150 feet by 30 feet. How many square meters is the garden?

18-119. A jet flies from New York to San Francisco, a distance of 2560 miles, in $4\frac{1}{2}$ hours. How many kilometers per hour does the jet fly?

18-120. Determine the cost of carpeting a room 8 meters by 4.5 meters at $14.50 per square yard.

18-121. How tall is a 6 foot 5 inch person in centimeters?

77 in = 195.58
× 2.54

18-122. What is the volume in kiloliters of a pool 50 feet by 25 feet and 6 feet deep?

18-123. Father Andy Marthaler drove 600 kilometers on 93.5 liters of gasoline on a Canadian fishing trip. Determine his average miles per gallon.

18-124. Five quarts of antifreeze are added to a 22-quart cooling system. How many liters of antifreeze would be needed for a 100-quart cooling system?

18-125. Water expands 9% when becoming ice. How large a cube of ice in cubic centimeters would you get from 0.57 kilogram of water?

THINK TIME

In time personal data will probably be known in metric measurement as well as English measurement. Complete your personal data in the space provided in Figure 18–16.

Measure	Type of Unit	Estimate	Actual
Height	centimeters		
Weight	kilograms		
Chest/bust	centimeters		
Waist	centimeters		
Hips	centimeters		
Foot	centimeters		

Figure 18–16   Personal data.

Estimate your measurements; this will help you relate measurement to the metric system. Try estimating and measuring other common objects that are listed in Figure 18–17. Add to the list.

Measure	Type of Unit	Estimate	Actual
Height of doorway	meters		
Weight of car	kilograms		
Distance to work	kilometers		
Amount of rainfall	centimeters		
Present temperature	Celsius		

Figure 18–17   Common objects.

PROCEDURES TO REMEMBER

1. To change from one unit to another in the metric system, move the decimal point the appropriate number of places.
   (a) To change a smaller unit to a larger unit, move the decimal point the appropriate number of places to the left.
   (b) To change a larger unit to a smaller unit, move the decimal point the appropriate number of places to the right.
2. To change from Celsius to Fahrenheit, multiply the Celsius degrees by 1.8, and add 32 degrees to the product; the result is degrees Fahrenheit: $F = 1.8C + 32°$.
3. To change from Fahrenheit to Celsius, subtract 32 degrees from the given Fahrenheit degrees, and multiply the difference by $\frac{5}{9}$; the result equals degrees Celsius: $C = \frac{5}{9}(F - 32°)$.
4. To convert from English to metric or metric to English, select the appropriate conversion fact and multiply by the conversion fact to obtain the new value.

CHAPTER SUMMARY

1. The seven basic units of measure in the SI metric system are meter (length), kilogram (mass or weight), kelvin (temperature), second (time), ampere (electricity), candela (luminous intensity), and mole (amount of substance).
2. Two common metric units that are not a part of the SI metric system are liter (capacity) and Celsius (temperature).
3. The interrelationship between volume, capacity, and weight in the metric system is

Volume	Capacity	Weight
1 cm^3 of water =	1 mL of water =	1 g of water
or	or	or
1 dm^3 of water =	1 L of water =	1 kg of water

Figure 18–18

4. Greek prefixes identify units *larger* than the base unit: *kilo* = 1000, *hecto* = 100, and *deka* = 10.
5. Latin prefixes identify units *smaller* than the base unit: *deci* = 0.1, *centi* = 0.01, and *milli* = 0.001.
6. Metric units, letter units, power of 10, and decimal units in the metric system are shown in Figure 18–2.
7. *Mass* is the quantity of material contained in a given body.
8. *Weight* is the measure of the force of gravity upon a given body.

CHAPTER TEST

T–18–1.   33 km = _____ m.

T–18–2.   1748 m = _____ km.

T–18–3. 5280 m = _____ km.

T–18–4. 31.5 cm = _____ m.

T–18–5. 84 sq in. = _____ cm².

T–18–6. 0.527 m² = _____ sq in.

T–18–7. 500 mL = _____ L.

T–18–8. 31.5 gallons = _____ L.

T–18–9. 21.5 L = _____ gallons.

T–18–10. 1500 g of water = _____ mL.

T–18–11. 44.2 dag = __.442_____ kg.

T–18–12. 53,000 kg = _____ metric tons.

T–18–13. 5000 mg = _____ g.

T–18–14. 35.3 dg = _____ g.

T–18–15. 26,400 ft = _____ km.

T–18–16. 100 ft = _____ m.

Chapter 18 / Fundamentals and Applications of the Metric System

T-18-17. 90°C = __194__ °F.

T-18-18. 95°F = _____ °C.

T-18-19. A house contains 4000 square feet of floor space. How many square meters is this?

T-18-20. A 12-millimeter wrench has an opening of how many thousandths of an inch?

T-18-21. How many inches in diameter is a 155-millimeter howitzer?

T-18-22. A mixture of 8 grams of salt and 8 grams of baking soda is poured into 225 milliliters of water. What is the total weight of the solution?

T-18-23. A radio station broadcasts at 100 megahertz. How many kilohertz is this?

T-18-24. At a cost of $12.50 per square meter, how much would it cost to blacktop a 50-foot by 16-foot driveway?

T-18-25. Two liters of concentrated liquid fertilizer are added to water to make a total of 10 gallons of plant food. How many milliliters of fertilizer are added per gallon?

INDEX

A

Acute angle, 170, 176
Addends, 3, 10
Addition, 4, 10, 19, 94, 97
Addition axiom, 103, 114
Adjacent side, 249
Algebra, 83–163
Algebraic expression, 85, 97, 102
Algebraic sum, 86
Altitude, 75, 79
Arc, 225, 238
Arc function, 277
Area, 121, 179–200, 181, 198
Arithmetic, 2–13
Axioms, 102

B

Bar, 86, 97
Base, 50, 51, 64, 75, 79, 97
Base number, 91
Binomial, 85, 97
Blueprints, 24
Brace, 86, 89, 97
Brackets, 86, 89, 97

C

Cancellation, 20
Capacity, 302, 306
Celsius, 298, 309, 316
Center, 186
Circles, 186, 197, 198
Circumference, 121, 181, 186, 198
Compass, 224, 238
Complementary angles, 170, 176
Cone, 206
Convert, 30, 40, 48
Cosine, 248
Cube, 79
Cylinder, 206

D

Decimals, 37–45, 51, 63
Decimal fraction, 45
Decimal point, 37, 41, 45
Decimal, repeating, 38, 41
Degrees, 170, 176
Denominator, 16–21, 31
Diameter, 186, 198, 225
Difference, 4, 10
Digits, 3
Discount, 60
Distance, 121
Dividend, 4, 10
Division, 4, 10, 94, 97
Division axiom, 106, 114
Divisor, 4, 10

E

End points, 168
Equal sign, 88, 97
Equation, 102–104, 114
Equilateral triangle, 168, 176
Evaluation, 108
Exponent, 69, 72, 79, 80, 85, 89, 91, 97
Exponential form, 38

F

Face, 152, 203, 220
Factors, 64, 72, 79, 85, 97
Formulas, 102, 114, 119, 127
    air/fuel, 141
    area, 121, 132, 197, 198
    average speed, 140
    circle, 135, 138
    circumference, 121
    cosine, 248, 255
    cube, 135, 138
    current flow, 138

Formulas (*Cont.*)
   cylinder, 142
   distance, 122
   electrical equivalent heat, 140
   ellipse, 129, 141
   expansion of gas, 141
   friction, 120
   geometric progression, 139
   Hooke's law, 159
   interest, 122
   latent heat, 140
   lateral surface area, 206
   magnetic intensity, 141
   mass/energy, 138
   parallel resistance, 140
   perimeter, 119, 133, 197, 198
   photographic enlargement, 141
   prismoidal, 141
   pulleys, 141
   sine, 241, 248
   sphere, 136, 138, 142
   tangent, 248, 255
   tap size, 141
   temperature conversion, 140
   thickness of pipe, 141
   thread depth, 124
   torus, 142
   total surface area, 204
   trapezoid, 124
   triangles, 124, 133, 137
   voltage drop, 139
Fractions:
   addition, 19, 30
   blueprints, 24
   common, 16–31, 42, 45, 51, 63
   complex, 16, 23, 31
   decimal, 37–44, 51
   division, 17, 31
   improper, 16, 30, 31
   multiplication, 17, 31
   proper, 17, 31
   subtraction, 21, 22, 31

G

Geometric constructions:
   angle, 226, 235, 237, 238
   circle, 225
   hexagon, 234, 238
   line segment, 225, 236, 237
   parallel lines, 230, 237
   perpendicular line, 229, 237
   square, 234, 237

Geometric constructions (*Cont.*)
   tangent, 231, 237
   triangle, 233, 237
Geometry, 165–239
   forms, 223–239
   plane, 168, 176, 165–239
   solid, 168, 176, 203–222
Greater than 89, 97
Grouping signs, 86, 88, 89, 97

H

Hectare, 300
Hemisphere, 209, 220
Hexagon, 175, 197, 238
Hypotenuse, 69, 79, 244, 255

I

Index of the root, 72, 79
Integers, 3, 10
Interest, 56, 64, 122
Interpolation, 262, 277
Invert, 17
Isosceles triangle, 168, 176

J

Jacob's Law, 63

K

Kelvin, 297, 309, 316
Kilogram, 297, 316
Kilogram table, 306

L

Lateral edge, 203, 220
Lateral face, 203, 220
Lateral surface area, 203, 206, 209, 220
Laws of Cosines, 267, 285–288, 292
Laws of Sines, 281–284, 292
Left member (side), 102, 114
Less than, 89, 97
Lever, 155
Like signs, 96
Like terms, 85, 97
Line, 168, 176

Line segment, 168, 176
Liter, 298, 316
Liter table, 302
Literal algebraic expression, 85
Literal coefficient, 85, 97
Literal equations, 111, 114
Literal quantities, 102, 114
Lowest common denominator, 19, 25, 30, 31

M

Mass, 305, 316
Means, 151
Meter, 297, 298
Meter table, 298
Metric equivalent tables, 311
Metric system, 38, 295–318
Minuend, 4, 10
Minutes, 170
Mixed numbers, 16, 30, 31
Monomial, 85, 97
Multiplicand, 4, 10
Multiplication, 4, 10, 17, 31, 94, 97
Multiplication axiom, 107, 114
Multiplier, 4, 10

N

Negative, 85, 86, 96, 97
Numerator, 16–23, 30, 31, 38
Numerical algebraic expression, 85, 97
Numerical coefficient, 85, 97

O

Oblique triangles, 279–292
Obtuse angle, 176
Octagon, 175
Opposite side, 247

P

Parallel, 229, 237
Parallel lines, 168
Parallelogram, 168, 176, 197
Parentheses, 86, 97
Pentagon, 175
Percent, 50, 64
Percent sign, 51, 64
Percentage, 50–56, 64

Perfect squares, 72, 79
Perimeters, 179–200, 181, 198
Perpendicular, 170, 229, 237
Pi, 186, 198
Place values, 38
Plane, 168, 176
Plane closed figure, 168, 176
Plane figure, 168, 176
Point, 38, 168, 176
Polygon, 168, 175, 176
Polynomial, 85, 97
Positive, 86, 96, 97
Power, 72, 79, 97
Powers of ten, 38, 69
Prime factors, 20, 30, 31
Prime number, 3, 10, 20
Principal, 56, 64
Prism, 203, 220
Product, 4, 10
Proportion, 153
Protractor, 173, 176
Pythagorean theorem, 253
Pyramid, 209, 220

Q

Quotient, 4, 10

R

Radian, 265
Radical sign, 72, 79
Radicand, 72, 79
Radius, 186, 198, 225, 238
Rate of percent, 50, 51, 64
Ratio, 150, 160
    inverse, 150, 160
    true, 160
Ratios, 148–163
Ray, 176
Reciprocal, 114
Rectangle, 168, 176, 197
Reduce, 17, 20, 23, 30, 31, 42, 51
Remainder, 30
Removal of signs, 88
Repeating decimal, 45
Rhombus, 168, 176
Right member (side), 102, 114
Right triangles, 75, 159, 168, 170, 176, 230, 234, 241, 243
Root, 103, 114

S

Scientific notation, 69, 79
Seconds, 170
Semicircles, 186, 197, 198
Septagon, 175
Sign of equality, 102
Sign of operation, 3, 10, 89, 97
Signs of quantity, 86, 97
Sine, 248
Solution of triangles, 243, 255
Solution set, 103, 114
Sphere, 123, 136, 209, 220
Square(d), 72, 79, 168, 176
Square root, 72–79
Straight angle, 170
Straight edge, 225, 238
Straight line, 168, 176
Subtraction, 3, 10, 21, 94, 97
Subtraction axiom, 103, 114
Subtrahend, 4, 10
Sum, 3, 10
Supplementary angle, 176
Systeme Internationale (SI), 297

T

Tangent, 229, 238, 248
Terms, 16, 31, 85, 97
Thread depth, 124

Total surface area, 204, 206, 209, 220
Transposition, 103, 114, 132, 145
Trapezoid, 124, 197
Triangle, 75, 79, 124, 168, 176, 197, 255
Trigonometric functions, 260, 261, 276
Trigonometry, 241–292
Trinomial, 85, 97

U

Unity, 19–23
Unlike signs, 86, 96
Unlike terms, 85, 97

V

Variables, 119
Vertex, 170, 176
Vinculum, 86, 97
Volume, 201–239, 212, 220, 302, 316

W

Weight, 305, 316

Z

Zero, 96, 97